Theoretical Biology

Series Editor

Yoh Iwasa, Kyushu University, Fukuoka, Japan

The "Theoretical Biology" series publishes volumes on all aspects of life sciences research for which a mathematical or computational approach can offer the appropriate methods to deepen our knowledge and insight.

Topics covered include: cell and molecular biology, genetics, developmental biology, evolutionary biology, behavior sciences, species diversity, population ecology, chronobiology, bioinformatics, immunology, neuroscience, agricultural science, and medicine.

The main focus of the series is on the biological phenomena whereas mathematics or informatics contribute the adequate tools. Target audience is researchers and graduate students in biology and other related areas who are interested in using mathematical techniques or computer simulations to understand biological processes and mathematicians who want to learn what are the questions biologists like to know using diverse mathematical tools.

More information about this series at https://link.springer.com/bookseries/15703

Tetsuji Tokihiro
Editor

Mathematical Modeling for Genes to Collective Cell Dynamics

Springer

Editor
Tetsuji Tokihiro
Graduate School of Mathematical Sciences
The University of Tokyo
Meguro-ku, Tokyo, Japan

ISSN 2522-0438 ISSN 2522-0446 (electronic)
Theoretical Biology
ISBN 978-981-16-7131-9 ISBN 978-981-16-7132-6 (eBook)
https://doi.org/10.1007/978-981-16-7132-6

This Springer imprint is published by the registered company Springer Nature Singapore Pte Ltd.
The registered company address is: 152 Beach Road, #21-01/04 Gateway East, Singapore 189721, Singapore

Preface

The human body is made up of tens of trillions of cells. Inside each cell, DNA, which carries genetic information, undergoes a process of transcription to produce the proteins that characterize the cell, and each cell acquires a variety of properties that are unique to each cell type. Cells also come together to form aggregates, acquire characteristic patterns and functions, and further form organs to carry out various biological activities. In processes that link DNA to organs, biodynamics, such as transcription by polymerases or formation of vascular networks by endothelial cells, play essential roles.

The main subject of the present volume is such biodynamics and its mathematical modeling. The topics are dynamics of RNA polymerases in transcription, construction of vascular networks in angiogenesis, and synchronization of cardiomyocytes. Statistical analysis of single-cell dynamics and classification of proteins by mathematical modeling are also presented. We show the experimental results and the mathematical models that can be used for the analysis of them. As we also give elementary introductions for the topics and mathematical approaches, the contents will be useful for both researchers in this field and interested beginners.

Recently, the importance of the collaboration between mathematics and biological sciences has been acknowledged, and new and fruitful outcomes have been appearing. We hope that the present volume gives good examples of the fruitful collaboration between mathematics and biological sciences.

On behalf of all the authors, I would like to express my sincere thanks to Professor Yoh Iwasa for giving us the opportunity to contribute to the *Theoretical Biology* series and for the helpful comments. Thanks are also due to our colleagues and friends, in particular the members of iBMath (Institute for Biology and Mathematics of Dynamic Cellular Processes), for various help and support. I also wish to thank Ms. Reshmi Rema and Ms. Fumiko Yamaguchi at the Springer Nature for their support and cooperation.

Tokyo, Japan
August 2021

Tetsuji Tokihiro

Contents

Chapter 1
Transcription Dynamics: Cellular Automaton Model of Polymerase Dynamics for Eukaryotes

Yoichi Nakata, Yoshihiro Ohta, and Youichiro Wada

1.1 Brief Review of Transcription

The cells of living organisms make proteins to reproduce themselves and express various functions in their activities. The making process of proteins is divided into two stages. First, a molecular motor called RNA polymerase (in eukaryotes, three types of RNA polymerases and RNAPII are in charge) recognizes an amino acid code described by deoxyribonucleic acid (DNA) and duplicates it using ribonucleic acid (RNA), resulting in the generation of messenger RNA (mRNA). This process is called transcription. Next, newly synthesized mRNA is decoded by ribosome, and a sequence of amino acid, namely protein, will be generated (Fig. 1.1).

To respond to both outer and inner stimuli, transcription of required protein coding gene will get started, terminates once the necessary amount of the proteins has been created, and the target protein depends on the cell type and situation. In the previous notion, genomic DNA was considered to move very flexibly during transcription process. This is because DNA is a linear polymer of alternating phosphates and sugars, each of which is bound to one of the four types of bases—adenine, cytosine, guanine, and thymine (Figs. 1.2 and 1.3). Triad of nucleic acid corresponds to a single amino acid, and this is called as codon. In humans, DNA consists of about 6 billion deoxyribose (i.e., it has 2^{12*10^9} bits of data). But only a tiny percentage of them are used [5].

Y. Nakata (✉) · Y. Wada
Isotope Science Center, The University of Tokyo, Tokyo, Japan
e-mail: ynakata@ric.u-tokyo.ac.jp; wada-y@lsbm.org

Y. Ohta
Arithmer Inc., Tokyo, Japan
e-mail: ohta@arithmer.co.jp

T. Tokihiro (ed.), *Mathematical Modeling for Genes to Collective Cell Dynamics*, Theoretical Biology, https://doi.org/10.1007/978-981-16-7132-6_1

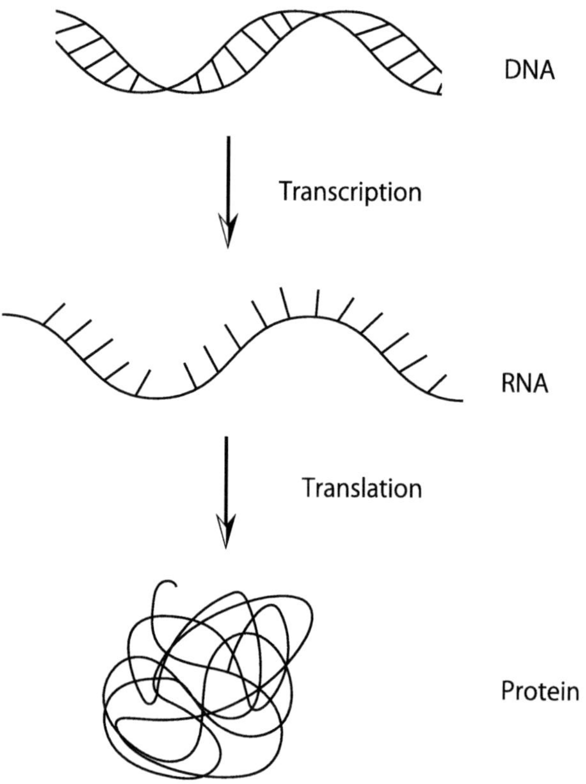

Fig. 1.1 Picture of central dogma. This picture is from https://www.genome.gov/about-genomics/fact-sheets/Transcriptome-Fact-Sheet

If we focus only on the base information, we do not determine which terminal should be selected as the start point, so it is necessary to determine the order. In the sugar, each carbon atom is labeled as in Fig. 1.2, and the 5' and 3' carbons are involved in polymerization. The terminal deoxyriboses have 5' or 3' carbon unused for polymerization, which is called 5'-terminal and 3'-terminal, respectively. The direction of the sequence is determined by the order from 5'- to 3'-terminal. This is also the orientation of the electrical polarity of DNA. However, it is noted that this orientation does not always coincide with the actual direction of RNAP reading DNA during transcription. In a single spice, genomic DNA has an identical sequence among all the cells and individuals (only immune response-related cells are exceptions). For chemical stability, the DNA forms a double helix structure by hydrogen bonding with the complementary strand determined by the complementary relationship of A-T and G-C. Therefore, DNA sites are counted by

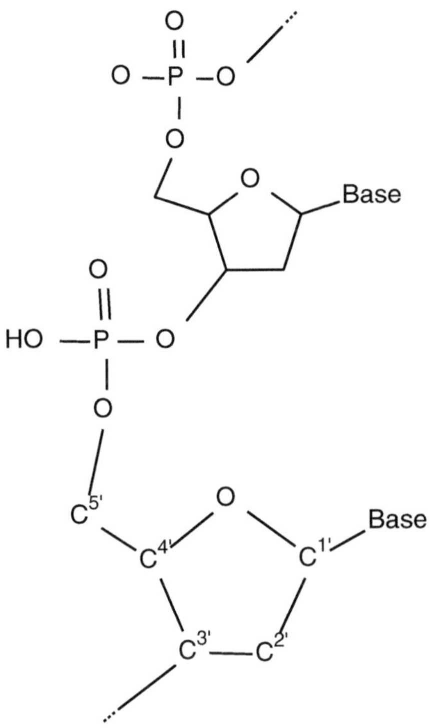

Fig. 1.2 Chemical structural formula of DNA

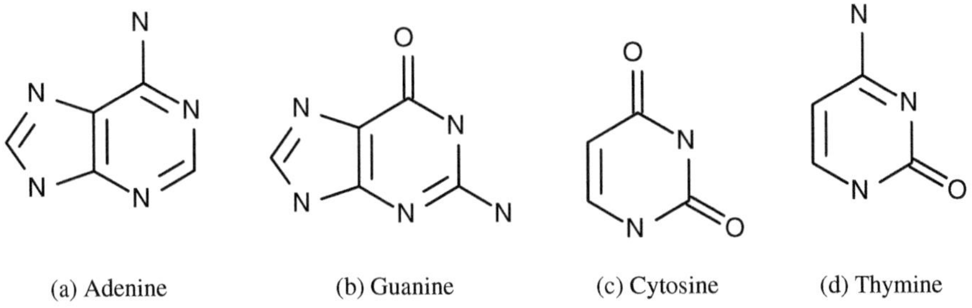

Fig. 1.3 Structures of four nucleotides

base pairs. The DNA double helix is stored in a form that wraps around protein octamers called histone. A set of this histone and the wrapping DNA is called a nucleosome, and a structure of compacted nucleosomes is called chromatin. Among chromatins, that of tightly bonded by nucleosomes are called heterochromatin regions, and conversely, that of loose bonds are called euchromatin regions. In transcription, euchromatin regions are prone to accept active RNAPIIs.

During cell division, chromatin gathers to form a pair of rods called chromosomes. The number of chromosomes varies depending on the organism, but in

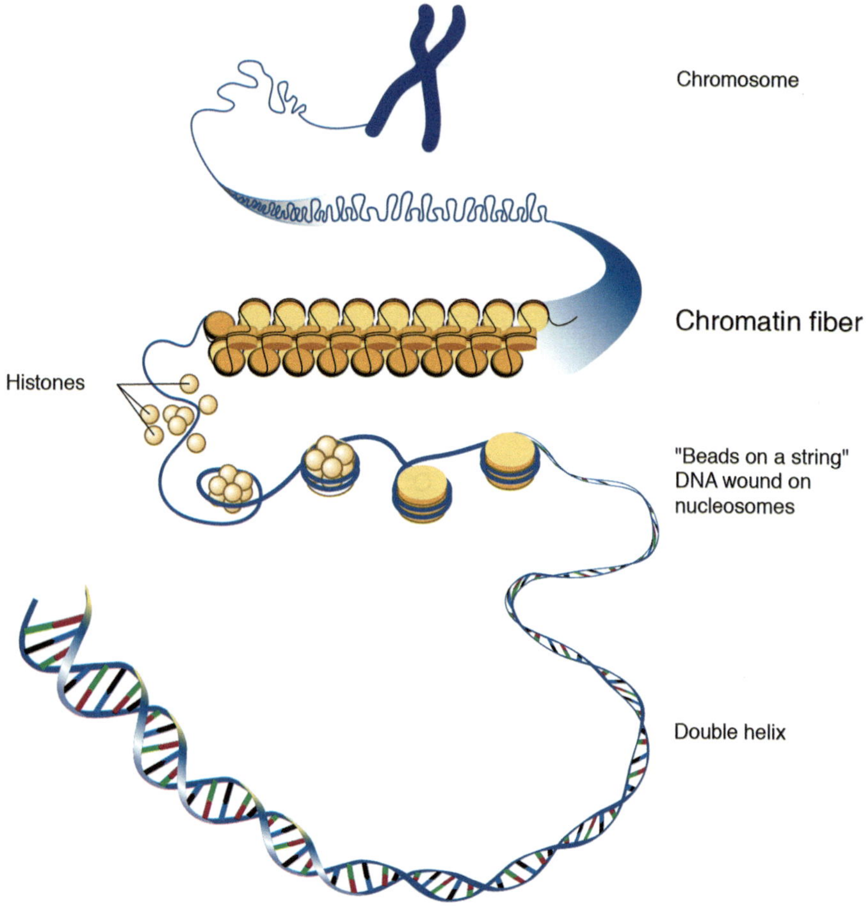

Fig. 1.4 A hierarchy of DNA. (This picture is a modified version of the one from https://www.genome.gov/about-genomics/fact-sheets/A-Brief-Guide-to-Genomics)

humans, it is known that there are typically 23 pairs of chromosomes. The position of a base pair in the sequence of DNA is expressed by using chromosomes (Fig. 1.4).

Only a few parts of DNA carry protein sequence code, and it is only 1.5% in humans. DNA regions involved in the production of specific proteins are called genes. Approximately, 26,000 genes are known in humans. Furthermore, in eukaryotes, there are also DNA regions within genes that are independent of the sequences of the proteins produced along the way. This region is called intron, and conversely, the region coding protein sequence is called exon. The location of genes, exon, and intron and the location of chromosomes are specified and databased.

RNAP generates RNA based on all the information in the gene region regardless of introns or exons, and this preliminary product is called pre-messenger RNA.[1] From this product, the parts derived from the intron regions are spliced out at some timing (in many cases, it is thought that RNAP reaches the exon region from the intron region). Eventually, the RNA part originated from the exon regions will be composed, and this is called messenger RNA (mRNA). This editing procedure is called splicing. Next, the mRNA is chemically treated at both ends so that it can exist outside the nucleus in order to generate proteins. This is where a straightforward question emerges. Why is there an inefficient part like intron? It seems to be good that genes hold only the minimum information necessary to make proteins, and RNAPII reads the only such codes to create mRNA. One hypothesis is that, when splicing, the intron-corresponding part and the exon part sandwiched between them are decomposed together to increase the final product variation. This phenomenon is called alternatively splicing, and the mRNA in which a part of the generated exon disappears is called a splice variant. Besides, there are research reports that the length of introns affects the rhythm of transcription, and as a result, the morphology of organisms is determined by regulating expression [16].

The information described using four types of sequences A, T, G, and C is generically called a genome. By 2003, the Human Genome Project revealed all human genome information. Therefore, anyone can access genome information from the database. At first, it was thought that all life phenomena could be known entirely if the genome was known, but it gradually recognized that there were phenomena that could not be explained by genome information alone.

For example, a queen bee breeds both a worker one and the next generation queen one, but there is no genomic difference between them. It is known that only those who have been given a special meal (royal jelly) in childhood can become queen bees. In addition, for humans born to mothers who became hungry during pregnancy, there was a statistically significant difference in the prevalence of lifestyle diseases when children grew up, depending on when hunger occurred during pregnancy. What is happening to their bodies? Several experiments reveal that transcription (and function expression) is affected by the replacement of a specific hydrogen group with a methyl group or an acetyl one in the base moiety part of the DNA site or histone tails (protein chains growing from each histone). For example, transcription is suppressed by cytosine methylation of DNA. When a specific one of the hydrogen group in the 27th amino acid residue in histone tail of H3 histone (which is known to be lysine) is replaced with a trimethyl group, DNA is firmly wrapped around a histone, making DNA hard to be involved in transcription. Conversely, methylation on the 27th amino acid of H3 histone is replaced with an acetyl group, DNA wrap is weakened, and transcription enhanced. Note that these modifications happen not in the genome itself because these modifications do not change DNA sequences. However, DNA modification pattern

[1] mRNA is that finally produces proteins. Other RNAs exist and are thought to play an important role in controlling transcriptional dynamics. However, that is not covered in this book.

is thought to be as crucial as genomic information because it significantly impacts transcription. Such a factor affecting transcription other than the genome sequence itself is called an epigenetic information, namely epigenome. All of the previous examples are regarded as epigenomic effects. It is known that the epigenome retains its information during cell division, and the fertilized egg inherits the parent epigenome, but most of the egg is reset just before the egg begins to differentiate into various cells. However, it has been found that some of the epigenome information is not reset by differentiation and is inherited to offspring as a result [23].

From here, let us consider the rough behavior of RNAP during transcription (RNAPII in the case of eukaryotes). This is a complex composed of 12 subunits in the nucleus and binds to a promoter region called TATA box motif (literally including a repeated TATA sequence), locating about 5 kbp upstream of the gene, and RNAP drifting in the nucleus binds to TATA box binding protein. Both proteins attract each other, and RNAP finally attaches to the DNA site and begins to move toward the transcription starting site. After arriving, it starts transcription and moves on the DNA track while generating pre-mRNA by reading the base information of DNA. This transcription direction depends on the gene and does not necessarily match the DNA direction. Sometimes RNAP may move in the opposite direction with DNA. RNAP terminates transcription when it arrives at the end of the gene region and desorbs. RNAP again drifts in the nucleus after it leaves the DNA and until it binds to the protein again. In this model, the movement of single RNAP is affected by the position of other RNAPs that run in front in addition to the genome sequence. This is very similar to the movement of a car on the road as described below. RNAP overtakes, slows down, and sometimes collides. This model works reasonably well for organisms with simple structures such as prokaryotes. For example, it has been confirmed that RNAP collides or slows down in prokaryotes [34, 35]. However, such a model can be applied to lives with simple structures like prokaryotes. It is known that the dynamics of RNAPII in eukaryotes, including humans, behave more complicated due to the reasons as follows. First, the genes of eukaryotes include not only protein-encoding regions (exon) but also introns. The pre-mRNA regions synthesized from intronic regions degrade rapidly and do not leave in the mRNA, which is the final product of transcription. Second, several proteins control the dynamics of RNAPII. Such proteins combine specific sites on DNA tracks and prevent RNAPII movement physically or help RNAPII attach to the transcription start site or attach RNAPII for starting transcription [14, 19]. There is a relation between a specific epigenome modification for some DNA sites and attachment of corresponding transcription factors [12, 17, 20, 21]. Finally, it is speculated that RNAPII does not exist alone, but a huge protein complex includes several RNAPII. The interactions that mediate this complex can also occur over long distances in the genome coordinates [5, 8–10, 29].

In order to create a model that describes the dynamics of RNAPII, we must take these into account. In the following sections, we will explain what experimental results have been reported on transcriptional dynamics and what dynamics we make up based on that.

1.2 Experimental Result

In this section, we will briefly explain what RNAPII behavior during transcription has been known by several experimental results before modeling [19, 33, 40]. To develop an RNAPII transcriptional dynamics model, we need to know the location of RNAPII. As one method, Wada and Ohta et al. investigated where and how much RNAPII was currently at some time by measuring the copy number of RNA fragments, which were recollected from living cells, labeled by fluorescent dyes, and quantified by 25mer nucleotide probes. This method is called a tiling array.

Formerly, we delivered Tumor Necrosis Factor-alpha (TNF-α), one of the inflammatory cytokines, to human umbilical vein endothelial cells (HUVECs) and observed the movement of the generation of nascent RNA by inflammation stimulation. Based on the previous experiment, we focused on long five genes, like the SAMD4A gene region for observation because it has good inflammation responses and sufficient length for observing RNAPII dynamics.

Visualization of nascent RNA generation was done in the specific genes at intervals of 7.5 min from 0 to 180 min after stimulation (Fig. 1.5). By analyzing the data, the following was obtained:

Transcription is not constantly performed after stimulation, and there are periods during which transcription is actively performed, such as waves. It takes about 15 min from stimulation to the first transcription.

At the transcription start point, constitutive active transcription happens, but many will be terminated before reaching to the end of genes. There is something like a checkpoint that shuts down RNAP transcription until a stimulus is activated to activate it.

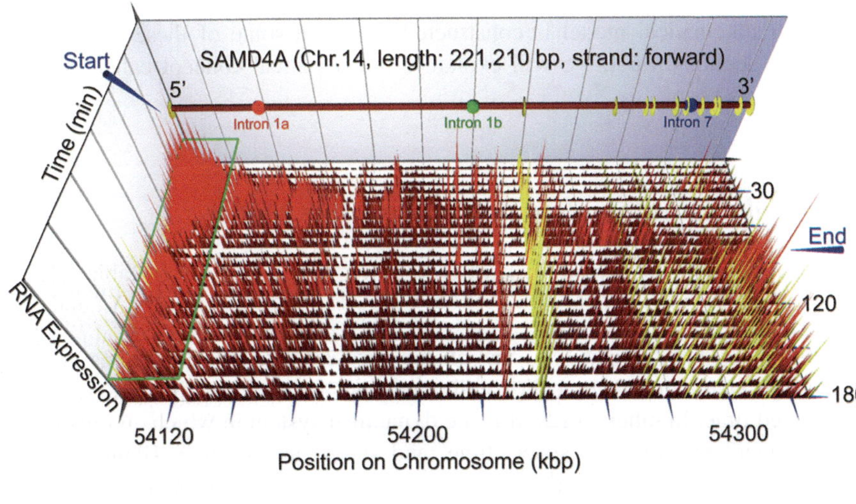

Fig. 1.5 Time evolution of the distribution of RNAPII on SAMD4A gene

When there is RNAPII performing transcription in the front, other RNAPIIs near the rear do not perform transcription.

By examining five representative genes, the transcriptional wave appears to be traveling at an average speed of approximately 3.1 kb per minute. It does not move at a uniform speed everywhere but depending on the region. For example, it seems to go faster in introns and slower in exons. This result has also been reported by Kolasinska-Zwierz et al. [19] and Schwartz et al. [33].

At the DNA site where a protein called CTCF is bound, the movement speed of RNAPII becomes slow. This is because CTCF might physically hinder RNAPII, this is why CTCF localizes the RNAPII migration range. When CTCF was knocked down, the movement of RNAPII was spread, and it is observed to happen at a place where transcription should not be expected. It is already known that the presence of CTCF is confirmed at the boundary of the active transcription region.

Looking at the RNA density profile, the density does not increase or decrease continuously from the beginning, but an isolated peak at a specific DNA site (around 0) appears over time. Most of those places are in exons, so it seems that it can be explained by the intron–exon speed ratio, but for that, you have to set a very high-speed ratio. The fact that the first intron does not have such a speed ratio cannot be ignored.

Recently, it should be noted that next generation sequencer allows us to detect the position efficiently by identifying the sequences of nucleotide fragments bound and recollected from specific proteins, including RNAP. The detected localization of RNAPII was consistent to that of transcription wave.

Further, the authors have reported that the observed "wave of transcription" was well explained by the presence of a complex called a transcription complex composed of RNAPII, which had been predicted to exist. The transcription wave was considered to be generated by the rapid change of the chromatin structure, which might be caused by transcription complex, the so-called transcription factory. Here, a mathematical model is constructed based on some of these facts in the following sections. First, we will explain a mathematical concept called cellular automaton as a tool.

1.3 Cellular Automaton and Traffic Flow Model

1.3.1 What Is Cellular Automaton?

A cellular automaton (CA) is one of the discrete dynamical systems. It has discretized states in lattices called cells and updates states of the cells for each discretized time. In other words, it is the dynamical system in which all dependent and independent variables (space, time, and states) are discrete. Ulam and von Neumann first propose the cellular automata concept to solve the self-reproduction problem of lives. The cellular automata represent complex nonlinear phenomena even if their update rules are simple.

The elementary cellular automaton (ECA) is a class of cellular automata with one-dimensionally configured cells with only two states (0 or 1), and the update of a cell is determined by itself, one before and after. That is, by denoting $t \in \mathbb{Z}_{\geq 0}$, $j \in I$ ($I \subset \mathbb{Z}$ is a discrete interval), and $u^t_j \in \{0, 1\}$ as the time, the site number and the state of site number j at time t, respectively, under the assumption of the initial condition $\{u^0_n\}_{n \in I}$ and the boundary condition, the time evolution of $\{u^t_j\}$ is expressed as $u^{t+1}_j = f(u^t_{j-1}, u^t_j, u^{t+1}_j)$ with a function $f : \{0, 1\}^3 \to \{0, 1\}$.

Since an ECA is regarded as a partial-difference equation, one needs to determine a proper boundary condition. Most of the cases, one assumes the Dirichlet boundary condition $u^t_{j'} = 0$ (j' is a boundary of I) or the periodic one $\exists N, u^t_{j+N} = u^t_j \; \forall t \geq 0$.

The amount of functions f (the update rules of ECA) is $2^{2^3} = 256$. We identify them by integers $k := \sum_{i,j,k=0,1} f(i, j, k) 2^{2^2 i + 2j + k}$. The ECA with integer index M is called ECA Rule M or ECA M. The behavior of ECA was actively analyzed, for example, by Wolfram [43]. There are several classifications for them, but the most common one is as follows [42]:

Class I: Uniform
Class II: Periodical
Class III: Chaotic
Class IV: Complex

The ECAs in classes III and IV show interesting behavior and sometimes exhibit characteristics of a real system.

An interesting pattern may be drawn when states of cellular automata are properly arranged at each time and space. For example, the time evolution of ECA Rule 90 with an initial state where only one point is 1 and the others are 0 is shown in Fig. 1.6. The time-space pattern is similar to a fractal pattern called a Sierpinski gasket.

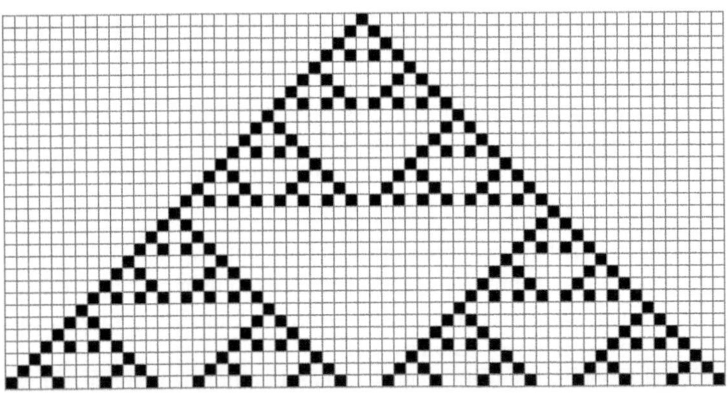

Fig. 1.6 An example of time-space pattern of ECA Rule 90. Black boxes mean $u^t_j = 1$ for these sites and white ones mean $u^t_j = 0$

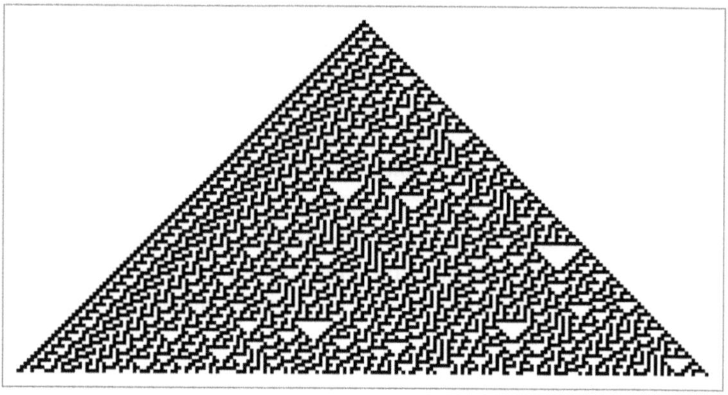

Fig. 1.7 An example of time-space pattern of ECA Rule 30. Triangle patterns seem to appear randomly

Figure 1.7 shows the space-time pattern of ECA Rule 30 with the same initial condition. It draws a similar fractal pattern but a more complex one. Interestingly, this pattern is very similar to that of a shell of the conidae. The question of why creatures create designs identical to ones in a simplified mathematical model is a fascinating subject, but it is beyond the scope of this book, so we will stop explaining here.

We can extend the elementary cellular automata by setting the cell configuration from a one-dimensional lattice to a higher-dimensional one or expanding the range of cells to be referenced on an update. We can express more complex natural phenomena for such extended models, such as solitons (the solitary waves that preserve their shapes after collision). One of the most famous examples is the Conway's life game, whose cells are configured on the two-dimensional lattice (with a proper boundary condition) and take two states—alive or dead. The state of a cell is updated using the following rules depending on the state of itself and those of its eight neighboring cells (Fig. 1.8):

- If the cell is alive, the next state is alive only if there are just two or three cells alive.
- If the cell is dead, the next state is alive only if there are just two cells alive.
- Otherwise, the next state of the cell is dead.

This rule indicates that life cannot survive alone but requires the cooperation of others. However, it also cannot survive if there are too many others around because they compete for limited resources. This exquisite balance of survival and death conditions creates a variety of patterns (Fig. 1.9).

Conway's life game (as well as most general cellular automata as a whole) is a good target for programming practice. It is a good study to write a program emulating the time evolution of the system by yourself. We will cite Golly (http://golly.sourceforge.net) as a ready-made program for the life game (Golly itself can

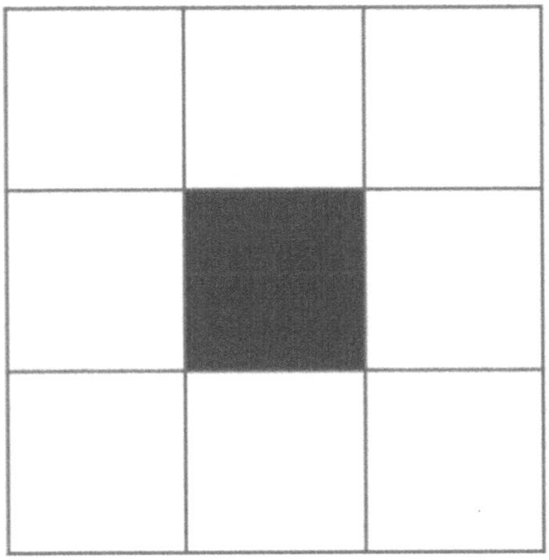

Fig. 1.8 The next state at the gray cell in the center is determined by eight surrounding neighbor sites in the figure

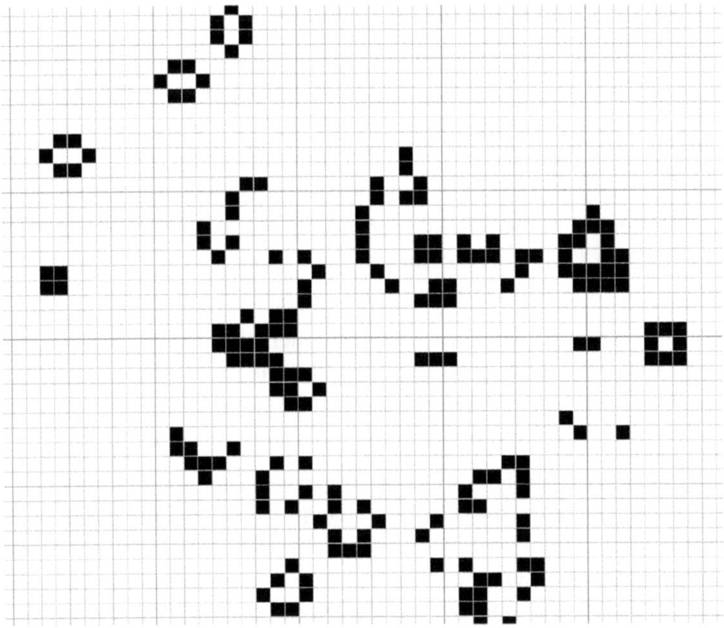

Fig. 1.9 A state of Conway's life game. The black cells mean alive and white cells mean dead

calculate time evolutions of more cellular automata). Anyway, when you have a program that describes the development rules for life games in some form, you can confirm that the system shows really fertile behavior when you start the simulation with appropriate initial values.

Due to this time evolution rule, the life game makes patterns that change periodically, translate while changing their shapes periodically, or regularly generate such parallel movement ones. By setting the initial state properly, one can express logic gates. Therefore, the life game can be a universal Turing machine.

Unlike the models using differential equations, it is very suitable with computer simulation because there is no discretization error or numerical error in simulation. It is possible to calculate quickly for improving calculation accuracy is unnecessary. Then, cellular automata models are often adopted to reproduce natural phenomena by computer simulations. One of the typical examples is the traffic flow model described in the next section.

Finally, the relationship between the phenomena by the CA and the similar ones by the differential equations is an interesting problem. Actually, this is one of the questions suggested by Wolfram. As one of the answers, it is known that we can derive the time evolution equation of a CA from a well-discretized differential equation by the limiting procedure called ultradiscretization [37].

1.3.2 Traffic Flow Cellular Automaton

Now, let us go back to ECA. We focus on ECA Rule 184. We note that there is no dependency of u^t_{j-1} to determine u^{t+1}_j when $u^t_j = 1$ and $u^{t+1}_j = 0$ when $u^t_{j-1} = u^t_j = u^t_{j+1} = 0$. By these facts, we can determine the update of the system even if it has an infinite amount of sites by setting an initial condition as $u^t_j = 0$ for $|j| \gg 1$.

Then, by interpreting that the sites are sequential boxes and each box has a ball if $u^t_j = 1$ for time t and j-th box and no balls if $u^t_j = 0$, we can rewrite the time evolution rule of ECA 184 as follows (Fig. 1.10):

- All balls are moving to the orientation that increases j.
- If there are no other balls at the next box, the ball moves there.
- If there is another ball at the next box, the ball stays.
- The next state is that by applying these rules above once for all sites.

Generally, the order to apply this update rule is very important, as such models. We apply this at the same time for all balls (remember the definition of the ECA update rule). That is, a ball cannot enter the next site simultaneously even if it is going to be empty by another ball leaving. This model is known as the simplest one to express traffic jams. The rule that one box can contain at most one ball corresponds to the exclusive volume effect. The traffic jam is finally solved when the number of balls (cars) is less than half of that of sites and never solved if more.

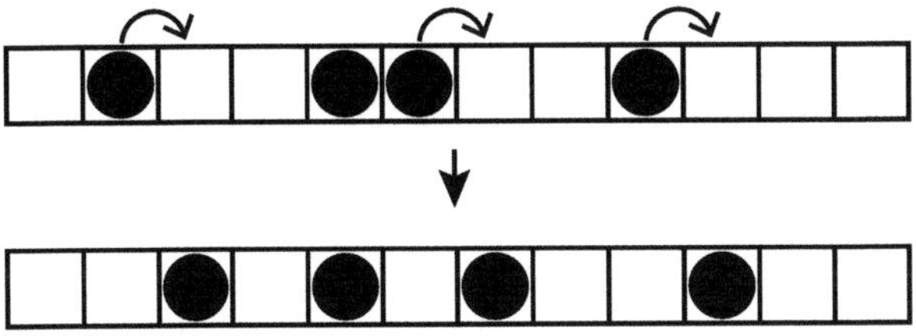

Fig. 1.10 Time evolution of ECA 184

There are several ways to express f explicitly, but the physically meaningful expression is, for example,

$$u_n^{t+1} = u_n^t + \min(1 - u_n^t, u_{n-1}^t) - \min(1 - u_{n+1}^t, u_n^t). \quad (1.1)$$

Transforming this equation, we have

$$u_n^{t+1} - u_n^t = \min(1 - u_n^t, u_{n-1}^t) - \min(1 - u_{n+1}^t, u_n^t). \quad (1.2)$$

Note that $\max(1 - u_n^t, u_{n-1}^t)$ expresses the number of moving particles from $n-1$ to n. The right-hand side means the total change of the number of balls at j. On the other hand, the left-hand side means the time change of the number of balls. Therefore, this expression is a CA analog of the equation of continuity in fluid dynamics. ECA 184 is obtained by the limiting procedure called "ultradiscretization" from a proper discretization of the Burgers' equation, which is a differential equation model expressing traffic jam (and originally proposed to represent shock waves in the compressed fluid) [25].

Now, we introduce a stochastic factor. Let p be a given parameter satisfying $(0 < p < 1)$. Each particle moves to the next site at probability p if possible. This model is called the Totally Asymmetric Simple Exclusion Process (TASEP). Recently, the mathematical model based on TASEP is often employed to solve traffic jam problems. By introducing stochastic i.i.d. variables $\{U_n^t\}_{t\geq 0, n\in\mathbb{Z}}$ that take 1 at probability p and 0 at $1-p$ and modifying the update rule (1.1), the time update rule of TASEP is expressed as

$$u_n^{t+1} = u_n^t + \max(1 - u_n^t, u_{n-1}^t, U_n^t) - \max(1 - u_{n+1}^t, u_n^t, U_{n+1}^t). \quad (1.3)$$

There are several ways to extend TASEP. For example, a ball can go to the second next site, or the moving probability depends on the distance of the next ball. By these extensions, we can have nearer expressions of the dynamics of real cars. For the extended CA models, a smooth traffic state becomes a metastable one when the number of cars is larger, and with a little perturbation, this state suddenly breaks, and the traffic jam finally occurs. This is considered as the mechanism of traffic jams occurring.

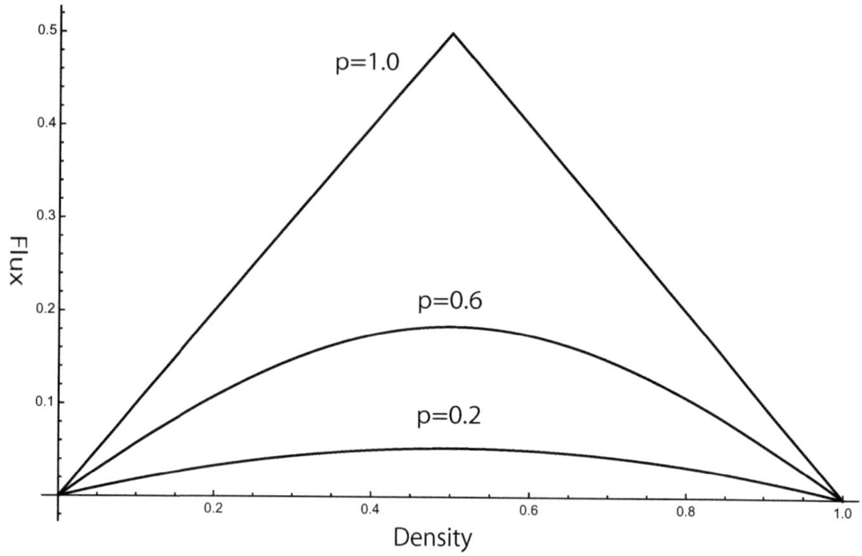

Fig. 1.11 Fundamental diagram of TASEP

Due to the observation above, the state of TASEP finally goes to a relaxed state. We define the flux of finite site TASEP as the average of moving balls per site (that is, $\sum \max(u^t_j, 1-u^t_{j+1})/N$) in the relaxed state. The graph of density of balls versus flux is called the "fundamental diagram," which is one of the most important pieces of information to understand the behavior of a model. For example, TASEP under the periodical boundary condition is expressed as Fig. 1.11. As referred before, if the number of balls is small, the traffic jam is finally solved, and all balls must move for each time evolution when sufficiently large time passed. Then, the flux is increasing monotonically depending on the density of particles. However, if the number of particles becomes larger, there are jams never solved. The number of jam trapped balls becomes larger as the density increased. Then, the flux is decreasing monotonically. In general, by denoting the number of sites and balls N and M, the density $\rho = M/N$ and the moving probability p and taking limit $M, N \to \infty$ to preserve M/N is constant, the flux is expressed as $\frac{1-\sqrt{1-4p\rho(1-\rho)}}{2}$ [30] and especially $\max(\rho, 1-\rho)$ as the limit $p \to 1$[25].

We can also set another boundary condition in which $j = 0$ and L are boundaries, and one injects a ball with probability α $(0 \le \alpha \le 1)$ at $j = 0$ if there are no balls, and one removes a ball with probability β $(0 \le \beta \le 1)$ at $j = L$ if there is a ball. Under this boundary condition, the system is known to have three phases of the flow—low density (LD), high density (HD), and flux maximized flow (MC).

There are two ways to explain the dynamics of TASEP: the view of the balls and the view of fields. Following the theory of fluid mechanics, we call the dynamics expression of the former viewpoint "Lagrangian representation" and the latter "Eulerian representation." In the infinite ECA Rule 184, denoting the position of k-th ball at time t x_k^t, one can obviously write down the time evolution rule $x_k^{t+1} = x_k^t + \min(1, x_{k+1}^t - x_k^t - 1)$. Here, it should be noted that the direction of movement is unique, and no particles overtake others. Due to these properties and the identity of the max calculation, we can directly relate these two dynamics representations [18].

1.4 Cellular Automaton Model of Transcription Dynamics

In this section, we introduce an RNAPII dynamics model by virtue of the experimental results described in section 1.2. Remember that RNAPII attaches DNA at the transcription starting site and starts transcription. Then, we coarse-grain target DNA track to a finite amount of sequential sites employ TASEP as a basic model for RNAPII dynamics on coarse-grained sites[31, 32].

We also set the stay time of RNAPII at each site [27]. The ball that arrived at a site has to stay during this time (of course, it still keeps staying after the time if there is another ball at the next site). This stay time is introduced to consider the difference of the velocity between introns and exons or due to the proteins that block RNAPII movement, for example, CTCF/cohesin (Fig. 1.12) [36].

At the given time, the ball is injected at the start site if it is empty. This means that the RNAPII free in the nucleus attaches to the start site and starts transcription.

We set a region including SAMD4A and its neighbor as the target region and the size of cell 35 bp. This size corresponds to the size of one RNAPII.

Before the numerical simulations, let us first imagine what phenomena can occur. Hereafter, for ease, we employ only the velocity difference between introns and exons and consider the semi-infinite system with the left side boundary. For such a system, we can explain the dynamics by introducing time τ_j^k as the time when k-th injected balls arrive at site j. (Note that let x_k^t be the position of k-th ball at time t, $x_k^t = j$, while $\tau_j^k \leq t \leq \tau_{j+1}^k - 1$.) Then, the dynamics is written in $\tau_{j+1}^k = \max(\tau_j^k + \gamma_j, \tau_{j+1}^{k-2} + \gamma_{j+1} + 2)$, where γ_j is the minimum stay time at site j. By using τ_j^k, the time of k-th ball spent for staying site j is $\tau_{j+1}^k - \tau_j^k$ and the time interval between when $k-1$-th ball arrives at site j and k-th ball arrives there is $\tau_j^k - \tau_j^{k-1}$. For the boundary condition, $\{\tau_0^k\}_{k\in\mathbb{N}}$ is given. For the ECA Rule 184, γ_j is identically equal to 1, which generates the trivial dynamics with no collision. To generate other states except for the free flow, one should employ the mechanism to stop the balls. In this system, the jam can occur when the latter ball at the intron site catches up with the former one at the first exon site due to the velocity gap.

Let us consider the dynamics only of $k-1$-th and k-th balls and their interval under the assumption that $k-1$-th ball never collides $k-2$-th ball since their interval is sufficiently long.

If these two balls are both in an intron and they are adjacent, the k-th one has to wait for one step until the next ball leaves. However, such a situation cannot happen unless all forward sites are occupied due to an existing jam. Therefore, under the

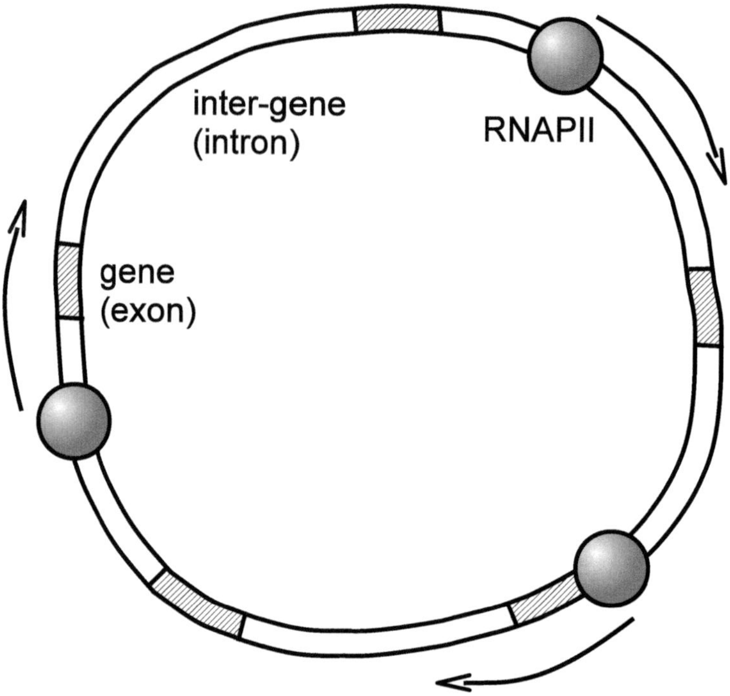

Fig. 1.12 Picture of the traffic flow model with staying time. Exon regions and intron regions appear alternatively and the staying time in exon regions is longer than that in intron ones

assumption, the balls can form only the free flow, and by setting $\tau_j^k - \tau_j^{k-1} = T$ at site j, we have $\tau_{j+1}^k - \tau_{j+1}^{k-1} = T$.

If two adjacent balls are both in an exon region, one should hold $\tau_{j+1}^{k-1} < \tau_j^k$ because the latter ball can move to the next site after the former one leaves there. Then, the $k-1$-th ball first spends the stay time, and k-th one can be adjacent, but the $k-1$-th one moves and creates a vacant space, while k-th one waits for its own waiting time. Then, it never happens that k-th ball cannot move because of $k-1$-th one. That means that there is only free flow in the exon regions. Especially, in the case that k-th ball moves the site immediately after $k-1$-th one leaves, that is, $\tau_{j+1}^{k-1} + 1 = \tau_j^k$ holds, we have $\tau_j^k - \tau_j^{k-1} = \gamma_e$. In the case that $k-1$-th is in an intron and k-th is in an exon, $k-1$-th moves first, and the site becomes vacant. Then, it forms a free flow. The jam can occur only in the case $k-1$-th is in an exon and k-th is in an intron. The time interval preserves if that in the first intron region $\tau_0^k - \tau_0^{k-1}$ is more than γ_e. If not, two balls collide and final time interval becomes γ_e and the k-th ball extra waiting time $\gamma_e = (\tau_0^k - \tau_0^{k-1})$ reduces the time interval of k-th and $k+1$-th ball $\tau_0^{k+1} - \tau_0^k$ to $\tau_0^{k+1} - \tau_0^k - \gamma_e + (\tau_0^k - \tau_0^{k-1}) = \tau_0^{k+1} - \tau_0^{k-1} - \gamma_e$. For such a case, we can determine that k-th and $k+1$-th balls collide with the signature of $\tau_0^{k+1} - \tau_0^k - (\tau_0^k - \tau_0^{k-1} - \gamma_e)$ instead of $\tau_0^{k+1} - \tau_0^k$. Generally, the condition that

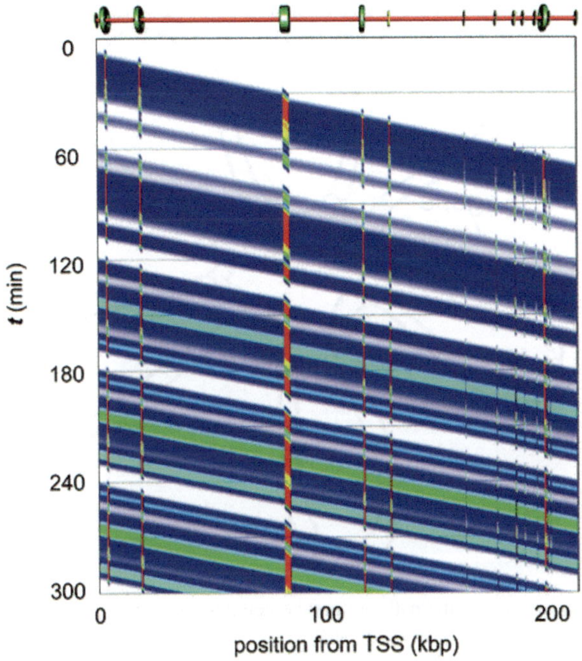

Fig. 1.13 Numerical simulation result of Ohta's cellular automaton model [27]. Colors express the density of RNAPII. Blue regions mean that they are not clouded, and red ones mean that they are clouded

the latter m balls get involved with traffics jams because k-th one overstays at the intron–exon boundary is written in $\tau_0^{k+m} - \tau_0^k - (m-1)\gamma_e$.

Due to the discussion above, if the number of particles is sufficiently small for the number of sites since the traffic jam can occur only in the boundary from the first intron and first exon and the balls after passing this boundary form the free flow, the traffic jam is finally solved.

We can apply similar discussion when there are three types of staying times. The bottleneck appears in the site with the maximal staying sites, and there are no factors that make jams after there.

In [27], the authors introduce periodic boundary condition because it is experimentally known that the transcription start site and the end site are spatially close at the active transcription regions and the RNAPII that finished transcription immediately attaches to the start site and starts transcription again. Figure 1.13 is the numerical simulation results, which match that discussed above because the target gene is sufficiently long.

Finally, we explain a model including RNAPII distribution gaps at introns. As described before, it cannot be explained only by the velocity difference between exons and introns. We have considered the system on the one-dimensional DNA track. However, since the real DNA is in the three-dimensional space and the track

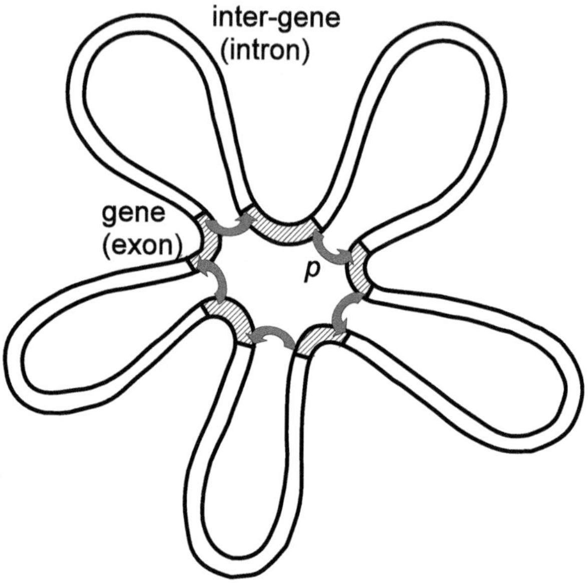

Fig. 1.14 Concept picture of the path-preference cellular automaton model. By being bound by CTCF, some regions of DNA form a ring shape. In such regions, transcriptionally active parts of the DNA, such as exons, are further bound to be spatially close to each other, and RNAPII is free to move through them by spatial diffusion effects

can bend as necessary, it can be possible that two distant sites on the DNA track are spatially close. Therefore, for such sites, one can consider that RNAPII shortcuts by virtue of the diffusion effects or through protein complexes [1].

The authors also proposed a model that enables RNAPII to jump from the end of an exon to the top of the next exon in each gene (Fig. 1.14) [28]. This effect can explain the splicing effects well because transcription products in skipped regions are not degraded but are ever not generated. Furthermore, by changing the target of jumping to the top of other exon regions, one can naturally explain the splicing variants.

The model is basically the same as that proposed before; that is, we consider the traffic flow cellular automaton model with finite amounts of balls and sites and the periodic boundary conditions. Each site is exon or intron. Due to the boundary condition, the number of exonic and intronic regions is the same, and we define the number K. We denote the top and the end of k-th exon as ι_{k-1} and ϵ_k, respectively (we consider the index by modulo K). Without loss of generality, we can set $\iota_K = 1$. We also assume that no more balls are injected after time evolution started in this system.

The basic time evolution rule of the balls is the same as ECA Rule 184, and we do not consider the velocity difference due to exon or intron regions and the proteins binding DNA sites. That is, in all sites, a ball moves to the next site if it is vacant.

Next, we also introduce the jump effect to this model. The ball at $j = \epsilon_k$ moves to $j = \iota_k$ with probability p and moves to the next site $j = \epsilon_k + 1$ by the ECA Rule 184 if the jump fails (probability $1 - p$ or there is another ball at $j = \iota_k$) and stays if there is another ball at the next site in addition to the jump failure. In general, we must first set the ordering to the destinations in the case that several balls are moving to the same site. In this model, we employ the prior for the jumping.

The authors also numerically simulated this model and plot distributions of RNAPII position after sufficient time passed and confirmed the distant peaks. They conclude that there is a diffusion effect because of DNA spatial closeness. Such closeness by RNAPII complexes is already proposed as the name of transcription factory [1, 5, 6, 15, 22, 41, 44].

Let us consider the simplest model with 1-exon and intron and $p = 1$ (that is, the ball at $j = \epsilon_k$ can jump to $j = \iota_k$ if it is vacant). In this model, we can observe a very interesting phenomenon. By calculating the fundamental diagram in the intron region, one finds a non-continuous gap as the number of balls is increasing, and this phenomenon is weakened by strengthening the stochastic effect. The path-preference model is a special case of traffic flow models with bifurcation and confluence. For such models, it is known that such gaps appear in the fundamental diagram [2]. In the path-preference model, we can obtain the exact flow after sufficient time passed by watching the trajectory [24].

In the exon region, only one pair of continuous vacant sites (we call this a *notch*) exists, and balls and empty sites appear alternately in other sites. Several clusters of balls and empty sites appear alternately exist in the intronic region, and other sites are vacant. In the case that there are one cluster and one notch, the system behaves time periodically and the flow is exactly obtained by

$$\langle J_{\mathrm{s}} \rangle = \frac{1}{T} \sum_{t=t_0}^{t_0+T-1} \frac{1}{N_{\mathrm{s}} - 1} \sum_{j=N_{\mathrm{m}}+1}^{N-1} f(j, j+1) = \frac{Q}{T} = \frac{2M - N_{\mathrm{m}} + 1}{2(2M + 1)}. \quad (1.4)$$

However, if the length of the cluster M is less than $N_{\mathrm{m}}/2$, the notch arrives at the end of exon before the top of the cluster arrives at the end of the intron. And the notch goes around in the exon again, which changes the periodic orbit pattern. Generally, the condition where the notch goes around exon γ times before the top of the cluster arrives at the end of the intron is written in

$$\frac{N_{\mathrm{s}} - (\gamma - 1)N_{\mathrm{m}}}{2} \leq M < \frac{N_{\mathrm{s}} - (\gamma - 2)N_{\mathrm{m}}}{2}, \quad (1.5)$$

and the flow is expressed as $\langle J \rangle = (2M - N_m + 1)/2/(2M + (\gamma - 1)N_m + 1)$. Surprisingly, under the condition that the number of balls satisfies, the flow takes the same values even if the details of the dynamics behave different. For example, if $\gamma = 2$, the time-space pattern of the system behaves like Fig. 1.15, in which there are one notch and two clusters. Therefore, we can conclude that there exist one notch and two clusters if the number of balls satisfies (1.5) because one of the

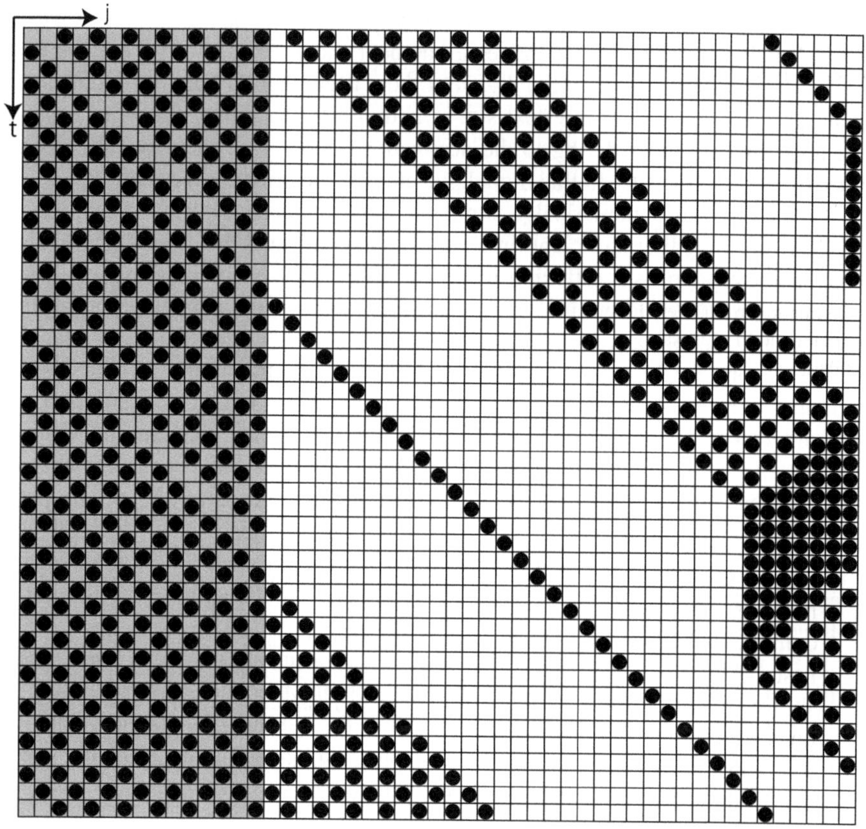

Fig. 1.15 A space-time pattern of the path-preference cellular automaton model. Two clusters of balls travel on the intronic region

clusters has no balls. Anyway, by observing a time-space pattern of periodic orbits, we can calculate the exact value of the flow.

We confirmed that there could be more complicated states with several notches with the same number of balls by setting initial states properly. Such states are established on delicate balances. Then they easily break down with little perturbation such as stochastic effects and reduce to the simpler states explained above.

In this section, we have explained the dynamics of the RNAPII cellular automaton model suggested by Ohta et al. Here, we note that there are other cellular automaton models to explain the dynamics of RNAPII, for example, [3, 4, 13, 26, 38, 39].

Recently, it is believed that the chromatin structure plays an important role in transcription and detecting chromatin structure to understand the transcription mechanics is actively studied. However, it is very difficult to directly observe the chromatin structure, which is the blob of the DNA chain and folding proteins.

The original purpose of authors' model is to explain to construct a model that explains the spatial proximity of distant sites on DNA coordinates during transcription through the movement of RNAPII with a cellular automaton method. However, it has also been found that this model is insufficient to grasp the spatial structure itself directly. One should adopt a more direct approach to capture a dynamical chromatin structure.

One of the major technologies to capture the chromatin structure is to hybridize a specific DNA site with fluorescence labeled probes (3D-FISH method). This approach can observe the structure directly and apply to living cells but only obtain spatial positions of some specific (hybridized) sites. Another method is to aggregate proteins with DNA fragments that contribute to the spatial connection of distant chromatin sites and to detect binding DNA sites from the sequences of the fragments [11]. This method can obtain the connection data of the whole genome, but the data indicate only the adjacency of two chromatin sites and the average of millions of cells due to the experimental method requirement. Therefore, one has to propose a physical or statistical model to guess the chromatin structure from the data [7] and then evaluates whether the estimated structure is correct by the 3D-FISH method.

References

1. Berg, O.G., Winter, R.B., von Hippel, P.H.: Diffusion-driven mechanisms of protein translocation on nucleic acids. 1. Models and theory. Biochemistry **20**, 6929–6948 (1981)
2. Brankov, J.G., Pesheva, N.C., Bunzarova, N.Z.: One-dimensional traffic flow models: Theory and computer simulations. https://arxiv.org/abs/0803.2625
3. Chowdhury, D., Santen, L., Schadsneider, A.: Statistical physics of vehicular traffic and some related systems. Phys. Rep. **329**, 199–329 (2000)
4. Chowdhury, D., Guttal, V., Nishinari, K., Schadschneider, A.: A cellular-automata model of flow in ant trails: Non-monotonic variation of speed with density. J. Phys. A Math. Gen. **35**, L573 (2002)
5. Cook, P.R.: Principles of Nuclear Structure and Function. Wiley, New York (2001)
6. Darzacq, X., Shav-Tal, Y., de Turris, V., Brody, Y., Shenoy, S.M., Phair, R.D., Singer, R.H.: In vivo dynamics of RNA polymerase II transcription. Nat. Struct. Mol. Biol. **14**, 796–806 (2007)
7. Dixon, J.R., Selvaraj, S., Yue, F., Kim, A., Li, Y., Y.Shen, Hu, M., Liu, J.S., Ren, B.: Topological domains in mammalian genomes identified by analysis of chromatin interactions. Nature **485**, 376–380 (2012)
8. Epshtein, V., Nudler, E.: Cooperation between RNA polymerase molecules in transcription elongation. Science **17**, 801–805 (2003)
9. Epshtein, V., Toulmé, F., Rahmouni, A.R., Borukhov, S., Nudler, E.: Transcription through the roadblocks: the role of RNA polymerase cooperation. EMBO J. **22**, 4719–4727 (2003)
10. Fraser, P., Bickmore, W.: Nuclear organization of the genome and the potential for gene regulation. Nature **447**, 413–417 (2007)
11. Fullwood, M.J., et al.: An oestrogen-receptor-α-bound human chromatin interactome. Nature **462**, 58—64 (2009)
12. Galburt, E.A., Grill, S.W., Wiedmann, A., Lubkowska, L., Choy, J., Nogales, E., Kashlev, M., Bustamante, C.: Structural basis of RNA polymerase II backtracking, arrest and reactivation. Nature **471**, 249–253 (2011)

13. Greulich, P., Garai, A., Nishinari, K., Schadschneider, A., Chowdhury, D.: Intracellular transport by single-headed kinesin KIF1A: effects of single-motor mechanochemistry and steric interactions. Phys. Rev. E **75**, 041905 (2007)
14. Guenther, M.G., Levine, S.S., Boyer, L.A., Jaenisch, R., Young, R.A.: A chromatin landmark and transcription initiation at most promoters in human cells. Cell **130**(1), 77–88 (2007)
15. Halford, S.E., Marko, J.F.: How do site- specific DNA- binding proteins find their targets? Nucl. Acids Res. **32**, 3040–3052 (2004)
16. Harima, Y., Takashima, Y., Ueda, Y., Ohtsuka, T., Kageyama, R.: Accelerating the tempo of the segmentation clock by reducing the number of introns in the hes7 gene. Cell Rep. **3**(1), 1–7 (2013)
17. Jin, J., Bai, L., Johnson, D.S., Fulbright, R.M., Kireeva, M.L., Kashlev, M., Wang, M.D.: Synergistic action of RNA polymerases in overcoming the nucleosomal barrier. Nat. Struct. Mol. Biol. **17**, 745–752 (2010)
18. Kanai, M., Nishinari, K., Tokihiro, T.: Stochastic optimal velocity model and its long-lived metastability. Phys. Rev. E **72**, 035102 (2005)
19. Kolasinska-Zwierz, P., Down, T., Latorre, I., Liu, T., Liu, X.S., Ahringer, J.: Differential chromatin marking of introns and expressed exons by H3K36me3. Nat. Genet. **41**, 376–381 (2009)
20. Kulaeva, O.I., Hsieh, F., Studitsky, V.M.: RNA polymerase complexes cooperate to relieve the nucleosomal barrier and evict histones. PNAS **25**(107), 11325–11330 (2010)
21. Li, B., Carey, M., Workman, J.L.: The role of chromatin during transcription. Cell **128**, 707–719(2007)
22. Li, G.W., Berg, O.G., Elf, J.: Effects of macromolecular crowding and DNA looping on gene regulation kinetics. Nat. Phys. **5**, 294—297 (2008)
23. Maekawa, T., Kim, S., Nakai, D., Makino, C., Takagi, T., Ogura, H., Yamada, K., Chatton, B., Ishii, S.: Social isolation stress induces ATF-7 phosphorylation and impairs silencing of the 5-HT 5B receptor gene. EMBO J. **29**(1), 196–208 (2010)
24. Nakata, Y., Ohta, Y., Ihara, S.: Periodic orbit analysis for the deterministic path-preference traffic flow cellular automaton. Japan J. Indust. Appl. Math. **36**, 25–51 (2019)
25. Nishinari, K., Takahashi, D.: Analytical properties of ultradiscrete Burgers' equation and rule-184 cellular automaton. J. Phys. A **31**, 5439–5450 (1998)
26. Nishinari, K., Okada, Y., Schadschneider, A., Chowdhury, D.: Intracellular transport of single-headed molecular motors KIF1A. Phys. Rev. Lett. **95**, 118101 (2005)
27. Ohta, Y., Kodama, T., Ihara, S.: Cellular-automaton model of the cooperative dynamics of RNA polymerase II during transcription in human cells. Phys. Rev. E **84**, 041922 (2011)
28. Ohta, Y., Nishiyama, A., et al.: Path-preference cellular-automaton model for traffic flow through transit points and its application to the transcription process in human cells. Phys. Rev. E **86**, 021918 (2012)
29. Papantonis, A., Larkin, J.D., Wada, Y., Ohta, Y., Ihara, S., Kodama, T., Cook, P.R.: Active RNA polymerases: mobile or immobile molecular machines? PLoS Biol. **8**, e1000419 (2010)
30. Schadschneider, A., Schreckenberg, M.: Cellular automaton models and traffic flow. J. Phys. A Math. Gen. **26**, L679 (1993)
31. Schmittmann, B., Zia, R.P.K.: Statistical Mechanics of Driven Diffusive Systems, vol. 17. Academic Press, New York (1995)
32. Schutz, G.M.: Exactly Solvable models for Many-Body Systems Far from Equilibrium, vol. 19, chap. 1. Academic Press, New York (2001)
33. Schwartz, S., Meshorer, E., Ast, G.: Chromatin organization marks exon-intron structure. Nat. Struct. Mol. Biol. **16**, 990–995 (2009)
34. Shearwin, K.E., Callen, B.P., Egan, J.B.: Transcriptional interference – a crash course. Trends Genet. **21**, 339–345 (2005)
35. Sneppen, K., Dodd, I.B., Shearwin, K.E., Palmer, A.C., Schubert, R.A., Callen, B.P., BarryEgan, J.: A mathematical model for transcriptional interference by RNA polymerase traffic in Escherichia coli. J. Mol. Biol. **346**, 399–409 (2005)
36. Sutherland, H., Buckmore, W.A.: Transcription factories: gene expression in unions? Nat. Rev. Genet. **10**, 457–466 (2009)

37. Tokihiro, T., Takahashi, D., Matsukidaira, J., Satsuma, J.: From soliton equations to integrable systems through limiting procedure. Phys. Rev. Lett. **76**, 3247 (1996)
38. Tripathi, T., Chowdhury, D.: Interacting RNA polymerase motors on a DNA track: effects of traffic congestion and intrinsic noise on RNA synthesis. Phys. Rev. E **77**, 011921 (2008)
39. Tripathi, T., Schülz, G.M., Chowdhury, D.: RNA polymerase motors: dwell time distribution, velocity and dynamical phases. J. Stat. Mech. P08018 (2009)
40. Wada, Y., Ohta, Y., S.I., et al.: A wave of nascent transcription on activated human genes. Proc. Natl. Acad. Sci. USA **106**(43), 18357–18361 (2009)
41. Wang, F., Greene, E.C.: Single-molecule studies of transcription: from one RNA polymerase at a time to the gene expression profile of a cell. J. Mol. Biol. **412**, 814–831 (2011)
42. Wolfram, S.: Theory and Applications of Cellular Automata. World Scientific Press, Singapore (1986)
43. Wolfram, S.: Cellular Automata and Complexity. Addison-Wesley, Boston (1994)
44. Yao, J., Ardehali, M.B., Fecko, C.J., Webb, W.W., Lis, J.T.: Intranuclear distribution and local dynamics of RNA polymerase II during transcription activation. Mol. Cell. **28**, 978–990 (2007)

Chapter 2
Angiogenesis: Dynamics of Endothelial Cells in Sprouting and Bifurcation

Hiroki Kurihara, Jun Mada, Tetsuji Tokihiro, Kazuo Tonami, Toshiyuki Ushijima, and Fumitaka Yura

2.1 Angiogenesis: A Biological Overview

2.1.1 Introduction

The establishment of systemic circulation is essential for the growth and maintenance of many multicellular organisms by providing oxygen and nutrients throughout the body. In vertebrates, systemic circulation is driven by the heart and blood vessels, which constitute closed and continuous luminal networks containing blood. While the heart acts as a pump with the systole–diastole cycle, blood vessels serve as a resilient conduit for circulating blood. Arteries and veins are usually comprised of three layers: tunica intima, tunica media, and tunica adventitia. Among them, the innermost tunica intima is comprised of a single-cell layer of endothelial cells (ECs). ECs also form capillaries between arteries and veins to serve as the site of

H. Kurihara · K. Tonami · T. Ushijima
Graduate School of Medicine and Faculty of Medicine, The University of Tokyo, Tokyo, Japan
e-mail: kuri-tky@umin.net; kazutonami@g.ecc.u-tokyo.ac.jp; ushijima-toshiyuki1123@g.ecc.u-tokyo.ac.jp

J. Mada
College of Industrial Technology, Nihon University, Chiba, Japan
e-mail: mada.jun@nihon-u.ac.jp

T. Tokihiro (✉)
Graduate School of Mathematical Sciences, The University of Tokyo, Tokyo, Japan
e-mail: toki@ms.u-tokyo.ac.jp

F. Yura
Department of Complex and Intelligent Systems, School of Systems Information Science, Future University Hakodate, Hokkaido, Japan
e-mail: yura@fun.ac.jp

T. Tokihiro (ed.), *Mathematical Modeling for Genes to Collective Cell Dynamics*, Theoretical Biology, https://doi.org/10.1007/978-981-16-7132-6_2

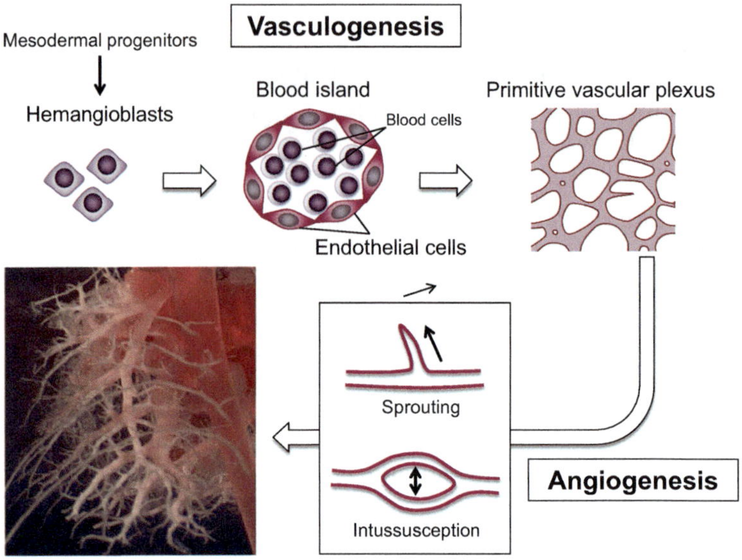

Fig. 2.1 Vasculogenesis and angiogenesis: two distinct processes to build vascular network in embryogenesis

gas and nutrient exchange. Thus, ECs are the only component of blood vessels that directly interacts with blood components.

Blood vessels are first formed as a luminal network of ECs during embryonic development via two distinct processes: vasculogenesis and angiogenesis[14, 45, 68] (Fig. 2.1). Vasculogenesis is a process in which ECs differentiate from mesodermal precursors in a previously avascular area to form a de novo vascular network. This event first occurs in blood islands of the yolk sac in association with primitive hematopoiesis, where mesoderm-derived hemangioblasts differentiate into ECs and red blood cells in the outer and inner layers, respectively. The outer endothelial sacs coalesce with each other to form the primitive vascular plexus. Vasculogenesis subsequently occurs in the embryonic body to produce two trunk axial vessels: the aorta and the posterior cardinal vein. Angiogenesis is the subsequent process to extend blood vessel networks by sprouting and intussusception from preexisting vessels. Following angiogenesis, newly formed vascular networks undergo remodeling, in which mural cells including smooth muscle cells and pericytes are recruited around a part of vessels to generate thick vessels and then arteriovenous connections are established. Besides embryonic development, angiogenesis occurs in response to tissue ischemia or increased oxygen demand during inflammation, wound healing, menstrual cycle, and so on. Tumor-associated angiogenesis is referred to as "tumor angiogenesis" and is a potential therapeutic target in many kinds of tumors [14, 30].

Hypoxia is among the major inducers of angiogenesis. When oxygen supply is decreased or oxygen demand is increased, the intracellular hypoxia-sensing mechanism is activated and hypoxia-inducible factor (HIF)-1α is accumulated in

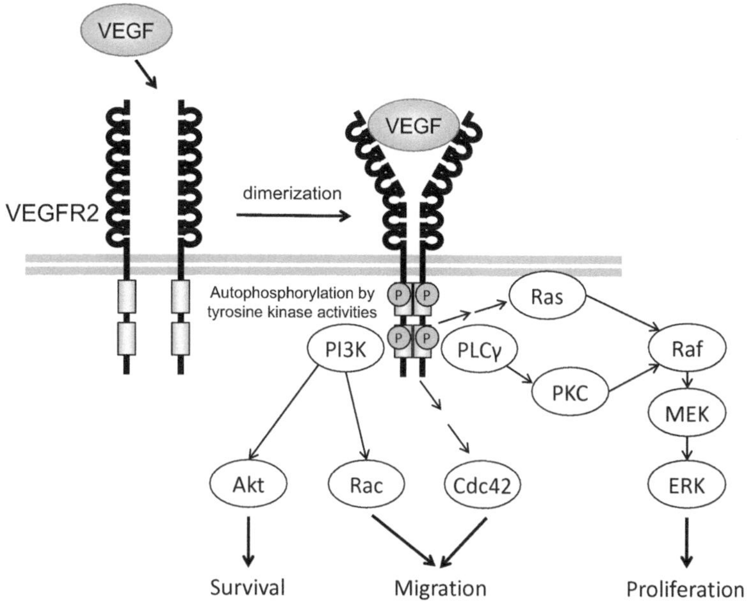

Fig. 2.2 VEGF–VEGFR2 signaling pathway

the cytoplasm, which is then translocated into the nucleus to induce transcriptional upregulation of vascular endothelial growth factor (VEGF), the most important angiogenic factor[27]. Several different members of the VEGF family bind three receptors with tyrosine kinase activity: VEGF receptor 1, 2, and 3 (VEGFR1, VEGFR2, and VEGFR3). The VEGF-A/VEGFR2 pathway plays a central role in not only physiological but also pathological angiogenesis. Binding of VEGF-A to VEGFR2 causes receptor homodimerization followed by autophosphorylation of tyrosine residues in the tyrosine kinase domain, which triggers the activation of mitogen-activated protein kinase (MAPK) and phosphoinositide-3-kinase (PI3K)-Akt/protein kinase B (PKB) pathways, leading to EC proliferation, migration, and survival[29, 66] (Fig. 2.2). Other factors that promote angiogenesis include fibroblast growth factors and angiopoietins, the receptors of which also possess the tyrosine kinase domain[47]. On the other hand, several factors such as angiostatin, endostatin, and thrombospondins exert inhibitory effects on angiogenesis[14, 45, 68]. The balance between these angiogenic promoters and inhibitors regulates EC migration toward avascular area and thereby the processes of angiogenesis in various situations.

2.1.2 Mechanisms Involved in Cell Migration

It might be worthwhile here to describe how cell migration is driven for readers unfamiliar to cell biology. In animals, many cells migrate by crawling on the sub-

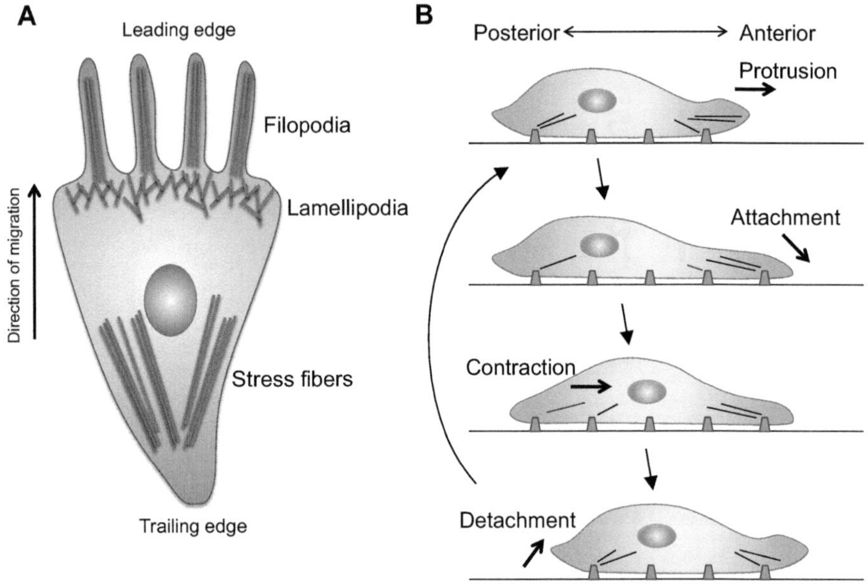

Fig. 2.3 Actin-based machinery of cell motility (**a**) and cyclic steps driving cell migration (**b**)

stratum (extracellular matrices) in response to various factors such as chemotactic substances. For example, leukocytes and macrophages migrate toward infection sites to attack pathogens as a part of the immune response. The process of migration can be dissected into several steps (Fig. 2.3a, b). (i) At the leading edge, two actin filament-dependent structures, filopodia and lamellipodia, are formed mainly by the activation of the Rho-family small GTP-binding proteins (G-proteins) Cdc42 and Rac1, respectively, and extend forward (protrusion). (ii) Then, the protruding actin filaments are anchored to the substratum through integrin-mediated focal adhesions (attachment). (iii) In the trailing cytoplasm, activation of RhoA, another member of the Rho-family small G-protein, leads to stress fiber formation and contraction of actomyosin, thereby drawing cytoplasm forward (traction). (iv) Focal adhesions at the trailing edge are then destabilized and detached from the substratum (detachment). These steps are integrated into the migration cycle (Fig. 2.3b), which takes place repeatedly to generate forward movement. For details, the readers are referred to Refs.[9, 18, 44].

2.1.3 Endothelial Cell Behaviors During Angiogenesis

From the viewpoint of cellular behaviors, angiogenesis can be regarded as a morphogenetic process driven by collective cell migration [17, 46]. When angiogenic stimuli such as VEGF-A act on the endothelial layer of preexisting vessels,

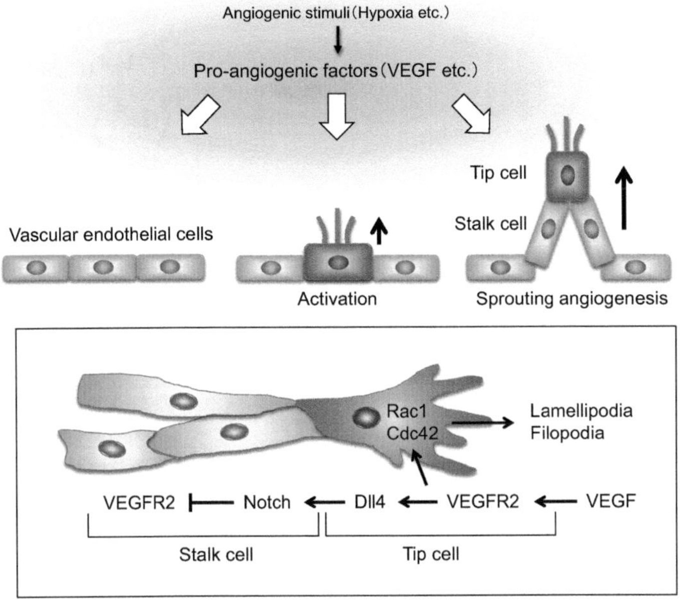

Fig. 2.4 Tip and stalk cell behavior during sprouting angiogenesis in response to angiogenic stimuli

selected cells respond to them by degrading the basal lamina and forming filopodial membrane protrusions to start chemotactic migration (Fig. 2.4). These cells act as *tip cells*, which have been supposed to lead angiogenic sprouting at the tip of elongating branches[23]. In addition to filopodial formation, VEGF-A activates the formation of lamellipodia through the VEGFR2–Rac1 pathway to promote migration[21, 53]. Tip cells further express the Notch receptor ligand Delta-like 4 (Dll4) through the VEGF-A/VEGFR2 signaling, which then acts on adjacent Notch-expressing cells to suppress tip cell activity and adopt stalk cell behavior trailing the tip cell in angiogenic sprouting[25, 37]. The dichotomous roles of tip and stalk cells are reminiscent of those of leader and follower cells in various types of collective cell migration, such as in the zebrafish lateral line primordium and Drosophila border cells [17, 46].

However, the reality is that the cellular behavior during angiogenesis is much more dynamic and complex. Two research groups including ours have reported characteristic EC movement visualized using time-lapse microscopic imaging[5, 28]. In in vitro angiogenesis models of mouse aortic ring assay[5] and embryoid body sprouts[28], ECs move with different speeds, forward and backward, and change their relative positional relationships to each other throughout sprouting and elongating branches (Fig. 2.5). Tip cells are constantly replaced by fast moving followers overtaking other ECs, resulting in a mixture of cell population constituting the elongating branches. This *cell-mixing* phenomenon was also confirmed in

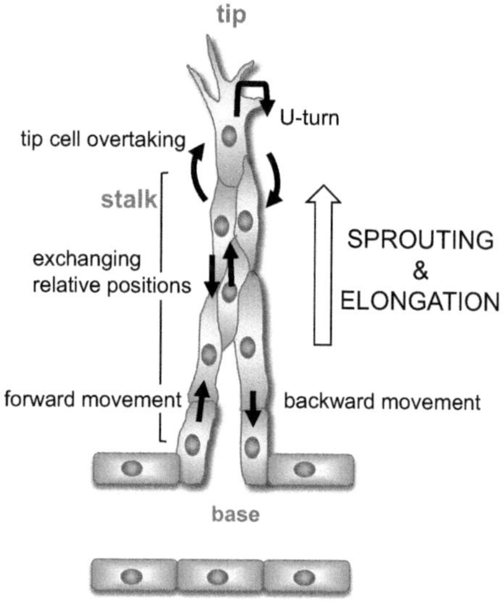

Fig. 2.5 Cell mixing phenomenon during sprouting angiogenesis

developing murine retinal vessels in vivo, indicating that it is physiologically relevant[5, 28].

In competition to occupy the tip cell position, ECs with high VEGFR2 and low VEGFR1 levels have been shown to be preferentially selected and the newly positioned tip cells express Dll4 in response to VEGF-A, which then re-establishes tip–stalk specification involving neighbors through the VEGFR-Dll4-Notch signaling circuit[28]. VEGF stimulates directed migration of tip cells without changing the frequency of overtaking, which is counteracted by Dll4-Notch signaling [5]. The directional elongation of EC sprouts is enhanced by the recruitment of mural cells, including smooth muscle cells and pericytes, through a Notch or other γ-secretase-dependent mechanism, which is likely coupled with decreased tip cell overtaking and retrograde movement[5].

As we have described, time-lapse live imaging has shown heterogenous, seemingly random EC movements during sprouting and branching morphogenesis. It is an open question how such complex multicellular behaviors are integrated into the formation of ordered architectures like blood vessels. One-dimensional agent-based mathematical modeling demonstrated that cell-autonomous EC movements, represented as a process driven by a simple stochastic transition among different states in terms of cell velocity, could recapitulate branch elongation[57]. However, tip cell behaviors could not be sufficiently explained by the cell-autonomous process alone. Instead, a coordinated mode of tip–follower interaction might be required for the forward tip motility. This prediction was biologically verified by different sets of experiments in in vitro aortic ring assay and in vivo zebrafish angiogenesis[57]. In addition, branching off from preexisting vessel-like structures is likely dependent on the regional cell density (our unpublished observation). These observations indicate that some deterministic processes based on intercellular interactions may be operating behind the random-appearing movements in branching morphogenesis.

2.1.4 *Cell-to-Cell Junction Proteins as a Modulator of Angiogenesis*

Intercellular interactions are primarily mediated by direct cell-to-cell contact through various junctions or the release of humoral factors. In epithelial tissues such as the vascular endothelium, cadherins, Ca2+-dependent cell adhesion molecules, form adherens junctions to stabilize the sheet structure and communicate with each other [42, 59]. Vascular endothelial (VE)-cadherin was first identified and characterized as an EC-specific member of the cadherin family, which is supposed to contribute to the vascular barrier function by constituting the endothelial adherens junction by homophilic interactions[33, 58, 61] (Fig. 2.6). The N-terminal region of VE-cadherin is comprised of five tandem extracellular cadherin (EC) domains, which mediate high-affinity homodimerization via trans-interaction and low-affinity assembly via cis-interaction in the adherens junction[10, 32, 63]. In the C-terminal intracellular region, VE-cadherin binds to p120- and α/β-catenins, which control cadherin trafficking and mediate attachment to actin filaments, respectively[32, 63].

VE-cadherin has proved indispensable for vascular development and morphogenesis, as revealed by experiments in which it was genetically inactivated in mouse embryonic stem (ES) cells[13, 24, 64]. VE-cadherin-deficient ECs failed to form vessel-like structures in ES cell-derived embryoid bodies[64]. VE-cadherin-deficient mice exhibited embryonic lethality with disorganized vasculature due to remodeling defects[13, 24]. Accompanying EC apoptosis indicates the role of VE-cadherin in endothelial survival[13]. On the contrary, a moderate decrease

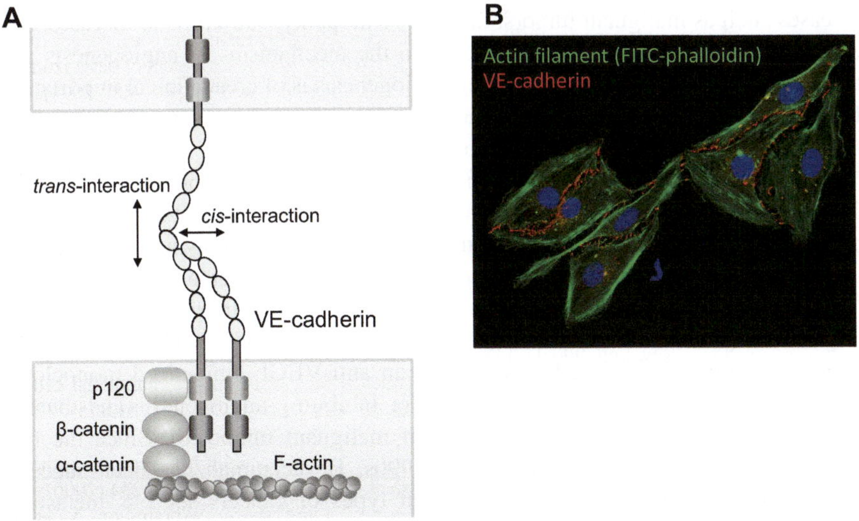

Fig. 2.6 Structure of VE-cadherin (**a**) and its localization in the endothelial cell adherens junction revealed by immunostaining (**b**)

in VE-cadherin expression results in hypersprouting due to failure to suppress VEGFR2/Rac1-dependent sprout formation[1, 19]. Thus, modulation of VE-cadherin function is critical for appropriate angiogenesis.

Crosstalk between VE-cadherin and VEGFR2 is well-known to play an important role in the regulation of angiogenesis. VE-cadherin associates with VEGFR2 to attenuate its downstream signaling through changes in the phosphorylation state[34, 35]. Conversely, VEGFR2/Rac1 signaling stimulates the phosphorylation of VE-cadherin by p21-activated kinase 1 (PAK1), which then recruits β-arrestin2 and promotes its clathrin-mediated internalization[22]. This β-arrestin2-dependent VE-cadherin endocytosis leads to increased endothelial permeability. In the context of angiogenesis, VEGFR2/Rac1 signaling causes a relative decrease in junctional VE-cadherin due to cell shape change, thereby triggering the formation of junction-associated intermittent lamellipodia (JAIL) under the guidance of the actin-nucleating WASP/WAVE/ARP2/3 complex[11]. Polarized JAIL formation is supposed to drive directed cell migration during sprouting angiogenesis. Such VE-cadherin/VEGFR2 crosstalk, also involving Dll4-Notch signaling in the tip–stalk interaction, is proposed to generate heterogeneity in EC dynamics, thereby driving angiogenic morphogenesis[8].

2.1.5 Clinical Significance of Angiogenesis

As mentioned before, angiogenesis not only contributes to the development of normal tissues but also plays significant roles in various (patho-)physiological states. For example, uncontrolled angiogenesis is often associated with a number of diseases such as malignant tumors, diabetic retinopathy, autoimmune diseases, and endometriosis. Therefore, investigation into the mechanisms of angiogenesis and therapeutic potential of agents targeting angiogenesis is of great clinical importance. In particular, tumor angiogenesis, which contributes to tumor growth, invasion, and metastasis, has attracted much attention[14, 30].

In many kinds of malignant tumors, VEGF expression is upregulated by HIF-1α, which is accumulated in the cytoplasm and translocated into the nucleus in response to hypoxia due to an imbalance between oxygen supply and consumption by excessively growing tumors[12, 14, 40]. As research on angiogenesis progressed, drug therapy aimed at inhibiting angiogenesis has appeared. In particular, molecular targeted drugs such as monoclonal antibodies have been developed and widely applied to clinical practice. Bevacizumab, an anti-VEGF humanized monoclonal antibody, is the first agent that was used for inhibiting angiogenesis and thereby suppressing proliferation and metastasis of malignant tumor[16]. Since the first clinical trial was conducted in the late 1990s, Bevacizumab has been successfully applied for the treatment of different types of cancers such as metastatic colorectal cancer[26], non-small-cell lung cancer[48], renal cancer[67], and breast cancer[41] in combination with standard chemotherapy. Thereafter, new drugs targeting the HIF–VEGF pathway, including VEGFR2-tyrosine kinase inhibitors, have been developed for antiangiogenic therapy and applied to various types of cancers[30, 36, 65]. Furthermore, combination of antiangiogenic therapy with immunotherapy with immune-checkpoint inhibitors, which has recently emerged as a promising therapeutic approach for the treatment of many cancers, has now opened a new possibility for cancer treatment[31]. Thus, researches on angiogenesis are of physiological and pathological significance, contributing to clinical practice in the treatment of various diseases.

2.2 Modeling by Differential and Difference Equations

In angiogenesis, networks of blood vessels are configured by elongation, bifurcation, and anastomosis of neogenetic vessels. An essence of these behavior is the collective motion of endothelial cells (ECs), and hence dynamics of the ECs in angiogenesis is very important to clarify this phenomenon. For this purpose, we often consider a simple model that represents significant features of the motion. If we model an EC as a particle, the particles interact with one another and will follow nonlinear equations of motion. The total system of ECs is expressed by simultaneous equations of motion, although they would be very different from those in Newtonian mechanics. In particular, the interactions between ECs are not spontaneous unlike Newtonian mechanics and we have to take into account the retardation effects. Then the equations of motion are expressed as difference equations or differential-difference equations with time delay. If the equations are linear, the qualitative features of solutions are not so different from the corresponding differential equations, but if they are nonlinear, we find that the solutions often exhibit quite different behavior.

Even if we can establish a plausible system of equations, it often becomes very difficult to analyze them even in numerical methods when the number of particles in the system becomes very large. Since we cannot treat incalculable number of ECs that constitute actual blood vessels, we often perform coarse-graining and consider dynamical equations for some macroscopic quantities such as density of ECs and average velocity. Then the system is expressed by partial differential equations of these quantities.

In this section, we briefly review several differential and difference equations that model typical biological phenomena. We explain the similarity and difference between continuous models and discrete models with simple biological phenomena. We do not consider the stochastic equations, which are explained in the next chapter. Some basic aspects of partial differential equation models are also presented.

2.2.1 Continuous Model and Discrete Model

In theoretical biology, one of the most fundamental processes is the death–birth process. The equation for this process is quite simple:

$$\frac{dy}{dt}(t) = ay(t), \tag{2.1}$$

where $y(t)$ is the population of some biological object at time t, and a is a constant. For example, $y(t)$ is a number (or density) of a bacteria in a petri dish, and a denotes its increasing rate. Solution to (2.1) is

$$y(t) = e^{at} y_0, \qquad (y(0) = y_0). \tag{2.2}$$

Hence the quantity y increases (decreases) exponentially for $a > 0$ ($a < 0$). The discrete version of (2.1) should be given by the following recurrence relation:

$$y_{n+1} = (1 + \alpha) y_n. \tag{2.3}$$

Here $\alpha = a\Delta t$ with the growth rate a and time interval Δt of one time step. Equation (2.3) is rewritten as

$$\frac{y_{n+1} - y_n}{\Delta t} = a y_n.$$

If we assume that there exists a smooth function $y(t)$ that satisfies $y(n\Delta t) = y_n$, then, by $\Delta t \to 0$, we obtain (2.1). Solution to (2.3) is

$$y_n = (1 + \alpha)^n y_0. \tag{2.4}$$

Equation (2.4) shows the exponential growth for $\alpha > 0$ and exponential decay for $-1 < \alpha < 0$. For $\alpha < -1$, the sequence (y_n) takes negative values and we need not consider this case for the discussion of population. Hence both (2.1) and (2.3) show essentially the same behavior and there is no qualitative difference.

Although exponential growth may take place in an early stage of population dynamics, it cannot continue indefinitely. There must be some adjustment to it. A plausible modification is to incorporate the population dependence in the growth rate.

$$\frac{dy}{dt}(t) = (a - by(t))\, y(t) \quad (a,\, b > 0). \tag{2.5}$$

The solution to (2.5) is given as

$$y(t) = \frac{a y_0}{b y_0 + (a - b y_0) e^{-at}}, \tag{2.6}$$

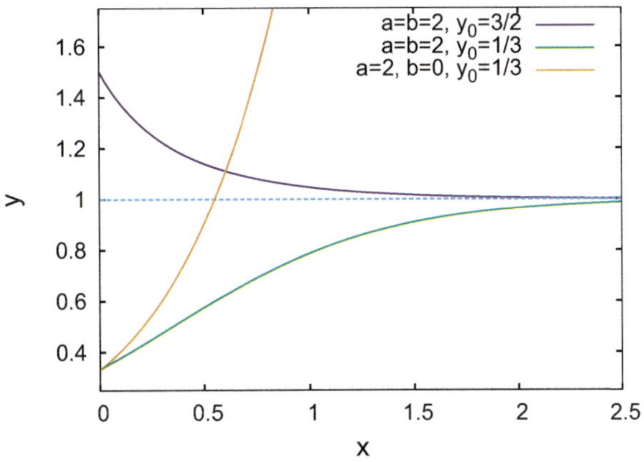

Fig. 2.7 Graphs of solutions (2.6)

for $y(0) = y_0$ (Fig. 2.7). A natural discretization of (2.5) is to use the forward difference as

$$\frac{y(t + \Delta t) - y(t)}{\Delta t} = (a - by(t))\, y(t). \tag{2.7}$$

By putting $t = n\Delta t$ and $y_n := y(n\Delta t)$, we have a recurrence equation:

$$\begin{aligned} y_{n+1} &= y_n + a\Delta t y_n - b\Delta t y_n^2 \\ &= \{1 + a\Delta t - b\Delta t y_n\}\, y_n \\ &= (1 + a\Delta t)\left\{1 - \frac{b\Delta t}{1 + a\Delta t} y_n\right\} y_n. \end{aligned}$$

When we rescale y_n as $x_n := \frac{b\Delta t}{1+a\Delta t} y_n$, we have

$$x_{n+1} = \alpha(1 - x_n)x_n. \tag{2.8}$$

The recurrence equation (2.8) is called the logistic map. For $0 < \alpha \le 4$, it maps the closed interval [0, 1] onto itself. For an initial data x_1 $(0 < x_1 < 1)$, which denotes initial population, the time evolution patterns strongly depend on the parameter α.

- $0 < \alpha \le 1$. The sequence (x_n) monotonically decreases to 0.
- $1 < \alpha \le 2$. The sequence (x_n) monotonically converges to $1 - \frac{1}{\alpha}$.
- $2 < \alpha \le 3$. The sequence (x_n) oscillates and converges to $1 - \frac{1}{\alpha}$.
- $3 < \alpha \le 1 + \sqrt{6}$. The sequence (x_n) approaches to a periodic oscillation with period 2.
- $1 + \sqrt{6} < \alpha < 3.56994567\ldots$. The sequence (x_n) approaches to a periodic oscillation with period 4, 8, 16, ... with increase of the parameter $alpha$. This behavior is an example of the so-called period-doubling.

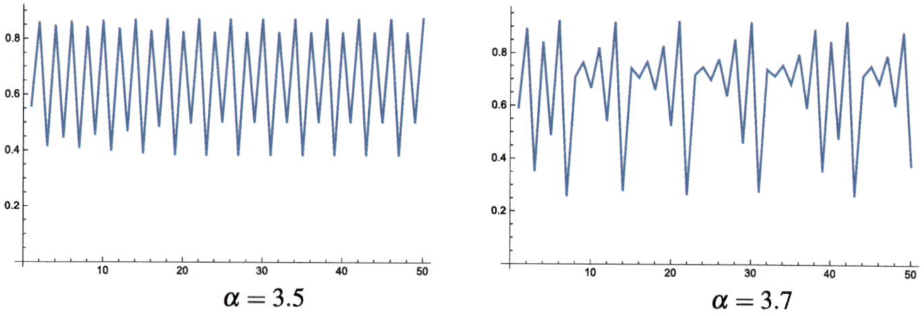

Fig. 2.8 Behaviors of the Logistic mapping (2.8) for $\alpha = 3.5$ and 3.7. The values x_n ($n = 1, 2, \ldots, 50$) are plotted with initial value $x_0 = 0.2$

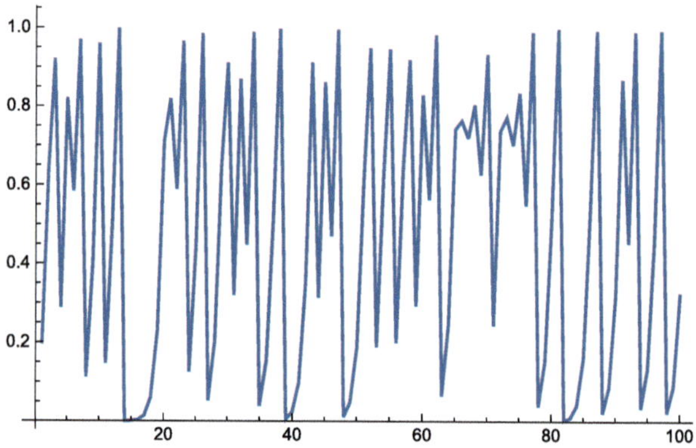

Fig. 2.9 The values x_n ($n = 1, 2, \ldots, 100$) of the Logistic mapping (2.8) for $\alpha = 4$ with the initial value $x_0 = 0.2$

- $3.56994567\ldots < \alpha \leq 4$. The sequence (x_n) shows chaos. For almost all initial conditions, it no longer approaches to oscillations of finite period. Slight variations in the initial data yield completely different results over time, which is the basic characteristic of chaos (Fig. 2.8).

The chaotic behavior is explicitly seen in the case of $\alpha = 4$. By putting $x_0 = \sin^2\theta_0$, we find that (Fig. 2.9)

$$\begin{aligned} x_1 &= 4\sin^2\theta_0(1 - \sin^2\theta_0) \\ &= \sin^2 2\theta_0 \\ x_2 &= \sin^2 4\theta_0 \\ x_3 &= \sin^2 8\theta_0 \\ &\cdots \\ x_n &= \sin^2 2^n\theta_0. \end{aligned}$$

These behaviors are very different from those of the continuous case (2.5). In general, discretization of nonlinear differential equations often changes the characteristics of their solutions although that of linear differential equations inherit their features. It should be noticed, however, that if we consider the following discretization instead of (2.7):

$$\frac{y(t+\Delta t)-y(t)}{\Delta t}=(a-by(t+\Delta t))\,y(t), \tag{2.9}$$

we have the recurrence relation:

$$x_{n+1}=\alpha x_n(1-x_{n+1}) \tag{2.10}$$

or equivalently

$$\frac{1}{x_{n+1}}=1+\frac{1}{\alpha x_n}. \tag{2.11}$$

This recurrence relation is easily solved and we have

$$x_n=\left(1-\frac{1}{\alpha}\right)\left\{\frac{x_0}{x_0+\alpha^{-n}(1-x_0-\alpha^{-1})}\right\}. \tag{2.12}$$

The sequence (x_n) monotonically converges to $1-\frac{1}{\alpha}$ for $\alpha > 1$ and to 0 for $0 < \alpha < 1$, just as the solutions to (2.5). Thus we see that the behavior of a nonlinear discrete dynamical system depends on how to discretize its original continuous equation. Nonlinear dynamical equations we encounter in the investigation of theoretical biology do not always have analytic solutions and we often employ numerical simulations. We should give attention to the relation of continuous equations to the discretized equations used in numerical simulations.

2.2.2 Newtonian Equation of Motion

One of the most well-known and important dynamical systems is that in Newtonian mechanics. Dynamics of particles is given by second-order differential equations. One particle dynamics is given as

$$m\frac{d^2\mathbf{x}}{dt^2}(t)=\mathbf{F}\left(\frac{d\mathbf{x}}{dt}(t),\mathbf{x}(t),t\right). \tag{2.13}$$

Here m is the mass of the particle, $\mathbf{x}(t)$ is the position vector of the particle, and $\mathbf{F}\left(\frac{d\mathbf{x}}{dt}(t), \mathbf{x}(t), t\right)$ is the force applied to the particle. A typical form of the external force $\mathbf{F}$ is

$$\mathbf{F}\left(\frac{d\mathbf{x}}{dt}(t), \mathbf{x}(t), t\right) = -\gamma\frac{d\mathbf{x}}{dt}(t) - \nabla V(\mathbf{x}(t)) + \mathbf{F}_o(t).$$

The parameter $\gamma > 0$ is the friction coefficient, and the first term on the right-hand side denotes friction by which the particle reduces its velocity. The function $V(\mathbf{x}(t))$ is called a potential such as the gravitational and the electric potentials, $\nabla := {}^t(\frac{\partial}{\partial x_1}, \frac{\partial}{\partial y}, \frac{\partial}{\partial z})$ denotes gradient, and the final term $\mathbf{F}_o(t)$ expresses the other external forces. When the potential is an isotropic harmonic potential with a coefficient $\kappa > 0$:

$$V(\mathbf{x}) = \frac{\kappa}{2}\|\mathbf{x}\|^2$$

and there is no other external force, the equation of motion is given as

$$m\frac{d^2\mathbf{x}}{dt^2}(t) = -\gamma\frac{d\mathbf{x}}{dt}(t) - \kappa\mathbf{x}(t).$$

Its general solution is given as

$$\mathbf{x}(t) = \begin{cases} \mathbf{A}_1 e^{-\lambda_1 t} + \mathbf{A}_2 e^{-\lambda_2 t} & (\gamma > 2m\sqrt{\kappa}) \\ \mathbf{B}_1 e^{-\lambda_0 t} + \mathbf{B}_2 t e^{-\lambda_0 t} & (\gamma = 2m\sqrt{\kappa}) \\ \mathbf{C}_1 e^{-\lambda_0 t}\cos\omega t + \mathbf{C}_2 e^{-\lambda_0 t}\sin\omega t & (\gamma < 2m\sqrt{\kappa}) \end{cases},$$

where

$$\lambda_0 = \frac{\gamma}{2m}, \quad \lambda_1 = \frac{\gamma + \sqrt{\gamma^2 - 4\kappa m^2}}{2m}, \quad \lambda_2 = \frac{\gamma - \sqrt{\gamma^2 - 4\kappa m^2}}{2m}, \quad \omega = \frac{\sqrt{4\kappa m^2 - \gamma^2}}{2m},$$

and the constant vectors $\mathbf{A}_1 \sim \mathbf{C}_2$ are determined by the initial conditions.

For N interacting particles, their motion is usually given by simultaneous differential equations as

$$m_i\frac{d^2\mathbf{x}_i}{dt^2} = -\gamma_i\frac{d\mathbf{x}_i}{dt} + \mathbf{F}_i + \sum_{j\neq i}\mathbf{F}_{ij} \quad (i = 1, 2, \ldots, N). \tag{2.14}$$

The first term on the right-hand side denotes the friction, the second term the external force similar to that in (2.13), and the last term the two-body interactions among the particles. For example, when the particles have electronic charge, $\mathbf{F}_{ij}$ has the following expression:

$$\mathbf{F}_{ij} = K \frac{q_i q_j}{\|\mathbf{x}_i - \mathbf{x}_j\|^2},$$

where q_i (q_j) is the electronic charge of the ith (jth) particle, and K is a constant determined by the permittivity of the medium. In most practical settings, the number of particles N is very large and Eq. (2.14) cannot be solved analytically or even numerically. If the system is in equilibrium, we can use principles of statistical mechanics; however, biological system is far from equilibrium and we need another principle. To establish such a principle is one of the most important subjects in theoretical biology, and substantial efforts are underway. We shall propose a mathematical model for the dynamics of endothelial cells in angiogenesis, which is a discrete analogue of the Newtonian equation with repulsive and attractive two-body interactions.

2.2.3 *Diffusion Equation*

Instead of treating numerous particles, we often introduce macroscopic field variables such as density or flux field. They mostly satisfy some partial differential equations or partial difference equations. Let us consider particles that move on d-dimensional square lattices.

A fundamental partial differential equation is the diffusion equation. For two spatial dimensions, the diffusion equation is given as

$$\frac{\partial u}{\partial t} = D\left(\frac{\partial^2 u}{\partial x^2} + \frac{\partial^2 u}{\partial y^2}\right), \tag{2.15}$$

where $u = u(x, y, t)$ is a density of some particle at position (x, y) and time t, and D is a positive constant called a diffusion constant. Equation (2.15) describes the collective motion of random walk particles in a plane. Let us consider random walk particles on a square lattice. The particle moves from a vertex to one of the four nearest vertices with equal provability at one time step. When we denote a position of a vertex by a pair of integers (l, m), a particle at the position (l, m) at time step n will move to $(l \pm 1, m \pm 1)$ at the next time step $n + 1$. Let the number of particles at vertex (l, m) at time step n be $U(l, m, n)$. If the number of particles is large enough, we have an approximate relation as

$$U(l, m, n+1) = \frac{1}{4}\left\{U(l+1, m, n) + U(l-1, m, n) + U(l, m+1, n) + U(l, m-1, n)\right\}.$$

When the lattice spacing is d and one time step is Δt, then the density of particles at the position $(x, y) := (ld, md)$ at $t = n\Delta t$ is given by $u(x, y, t) = U(l, m, n)/d^2$; hence, we have

$$u(x, y, t + \Delta t) = \frac{1}{4}\{u(x + d, y, t) + u(x - d, y, t) + u(x, y + d, t) + u(x, y - d, t)\}. \tag{2.16}$$

If $u(x, y, t)$ is smooth in time and space, we can expand in Taylor series as

$$u(x, y, t + \Delta t) = u(x, y, t) + (\Delta t)\frac{\partial u}{\partial t}(x, y, t) + O((\Delta t)^2)$$

$$u(x \pm d, y, t) = u(x, y, t) \pm d\frac{\partial u}{\partial x}(x, y, t) + \frac{d^2}{2!}\frac{\partial^2 u}{\partial x^2}(x, y, t) \pm \frac{d^3}{3!}\frac{\partial^3 u}{\partial x^3}(x, y, t) + O(d^4)$$

$$u(x, y \pm d, t) = u(x, y, t) \pm d\frac{\partial u}{\partial y}(x, y, t) + \frac{d^2}{2!}\frac{\partial^2 u}{\partial y^2}(x, y, t) \pm \frac{d^3}{3!}\frac{\partial^3 u}{\partial y^3}(x, y, t) + O(d^4).$$

Hence, from (2.16), we have

$$\frac{\partial u}{\partial t}(x, y, t) = \frac{d^2}{\Delta t}\left(\frac{\partial^2 u}{\partial x^2} + \frac{\partial^2 u}{\partial y^2}\right) + O(\Delta t) + O(d^4/(\Delta t)).$$

By fixing $\frac{d^2}{\Delta t} =: D$ and taking limits $d, \Delta t \to 0$, we obtain the two-dimensional diffusion equation (2.15). In the next section, we shall see that continuous limit of a discrete model for angiogenesis is described by a nonlinear diffusion equation that has a density-dependent diffusion coefficient.

2.3 Mathematical Model for the Dynamics of Endothelial Cells in Angiogenesis

2.3.1 Introduction

A lot of models for angiogenesis have been and is being constructed with various mathematical methods from different point of view [2, 6, 7, 51, 62]. For example, a hybrid model of cell migration on an elastic matrix of fibers[55], a mesoscopic lattice-based stochastic model in relation to deterministic continuous models[54], and a phase-field continuous model of sprouting angiogenesis described by compact partial differential equations[49] were proposed recently. Most models, single-cell-based models[3] in particular, mainly concern the pattern formation of vessel networks where chemoattractant gradient and chemotaxis of ECs play essential roles [15, 20, 50]. As for the experimental results in the dynamics of ECs along an elongating branch, which are shown in the previous sections, a stochastic four-state model and the corresponding differential equation model have been discussed[56]. In this section, we consider a simple mathematical model to explain the dynamics of ECs in the early stage of angiogenesis and show that it reproduces the cell-mixing behavior, elongation, and bifurcation of vessels without chemotaxis because no gradient of angiogenic factors such as vascular endothelial growth factor (VEGF) is required for branching morphogenesis in the experimental settings stated in the previous section. In the modeling angiogenesis, it is practically impossible to take all the factors into account even for its early stage of sprouting vessels, because it is quite a complex multicellular process; we have to consider individual and heterogenous shapes of ECs, interactions with extracellular organic materials, variety of chemical and biological reactions, time delay to the responses, fluctuations due to the environment of ECs, and so on. Instead we focus on the effects of the cell-to-cell interaction that we conjecture essential in the cell-mixing phenomenon, elongation and bifurcation, and show that deterministic two-body interaction between ECs plays an essential role in the cell-mixing effect. First, based on the investigation reported in [39], we consider a simple discrete dynamical system consisting of particles and paths that represent ECs and primary vessels, respectively. The particles with two-body interaction are supposed to move along the paths. Since interactions between particles are not instantaneous because they are caused by biochemical reactions with retardation effects, the positions and velocities of the particles are updated in discrete time steps in our mathematical model. Elongation and bifurcation of paths are determined by certain rules determined by the density and pressure of ECs at a tip. We do not incorporate stochastic fluctuation into the EC dynamics here. Stochastic fluctuation in EC dynamics may have some importance in quantitative analysis and it is straightforward to take it into our model. A continuous mathematical model described by differential equations, which was reported in [38], is also shown as an approximation of the discrete dynamical model.

2.3.2 *Discrete Model for EC Dynamics*

As described in the previous section (Sect. 2.1), we model the dynamics of ECs in sprouting blood vessels from murine aortic rings embedded and cultured in collagen gel placed in a shallow petri dish. The system is essentially two-dimensional, and the neogenetic vessels (primary blood vessels) of a branch are supposed not to have lumen yet and not to carry blood nor lymph. We neglect the effects of anastomosis (reconstruction of vessels) and cell division of ECs in neogenetic vessels. Actually cell division is rarely observed in the time span of the experiment (~5% a day), although it becomes nonnegligible as the vascular network develops considerably. We estimate its effect on vascular networks in the next subsection with an analytic continuous model. ECs are supplied from the basal sheet around the aortic explants, where cells seem to be relatively active in proliferation. Hence the vascular network may be treated as a tree graph on a plane that develops in time, and the ECs are regarded as particles immigrating on the graph. The particles are supposed to be supplied from the origin of the tree graph. The actual vessels, however, have spatial structures; they are wide near the root and thin at the tip. In the present model, we consider that the density of particles positively correlates with width of a vessel, that is, if the density of ECs in a branch is high, then the corresponding vessel is considered to be thick.

Let us consider the formation of one of the newly generated blood vessel networks that arise from the aortic ring and discuss the dynamics of ECs that also describes elongation and bifurcation of the vessel. We suppose that there is no neogenetic vessel around the aortic ring at the initial time $t = 0$ and a new blood vessel sprouts in a certain direction according to the supply of ECs for $t > 0$. The nth ($n \in \mathbb{Z}_{\geq 0}$) EC comes to the origin of this neogenetic vessel network at time step $t = \sum_{i=1}^{n} a_i$ with an initial velocity $v_{ini}(n) \geq 0$, where $a_i \in \mathbb{Z}_{+}$ is the time interval between the incidence of ith EC and that of $(i-1)$th EC. For the time being, we consider the dynamics of ECs only in the first neogenetic vessel and concentrate on the effects of two-body interactions between ECs in a branch. Elongation and bifurcation of vessels will be discussed in the next subsection. Although interactions among ECs are quite complicated and have not been well-elucidated, it is plausible to consider that direct contact of their membranes gives rise to the activation of chemical channels or mechanical driving force of ECs. For short distance, the interaction force will turn out to be repulsive due to excluded volume effect, while it will turn out to be attractive if the distance becomes larger because of the interaction with pseudopodia. Thus the mathematical model we investigate here is given as

$$x_n^{t+1} = x_n^t + v_n^t \tag{2.17}$$

$$v_n^{t+1} - v_n^t = -\gamma v_n^t + \sum_{k \neq n} F(x_n^t - x_k^t), \tag{2.18}$$

where $x_n^t \geq 0$ is the position of the nth EC measured from the origin at time step $t \in \mathbb{Z}_{\geq 0}$, the time unit ($\Delta t \equiv 1$) may correspond to the specific response time, the parameter γ ($0 < \gamma < 1$) denotes the coefficient of conflict, and F denotes the two-body interaction. For the function F, which denotes the two-body interaction between ECs, we adopt the following simple form:

$$F(x) := \begin{cases} \operatorname{sgn}(x) f_r & (0 < |x| \leq R_r) \\ -\operatorname{sgn}(x) f_e & (R_r < |x| \leq R_e) \\ -\operatorname{sgn}(x) f_a & (R_e < |x| \leq R_a) \\ 0 & (R_a < |x|) \end{cases}, \tag{2.19}$$

where $\operatorname{sgn}(x) := \frac{x}{|x|}$ and f_r, f_e, f_a are the positive constants for interaction strengths (Fig. 2.10).

Equation (2.17) means that v_n^t is the velocity of the nth EC at time step t, while (2.18) is the discrete analogue of the Newtonian equation of motion. The meanings of the parameters in the function F are as follows. The excluded volume effect is supposed to be dominant if two ECs approach within the typical size of a

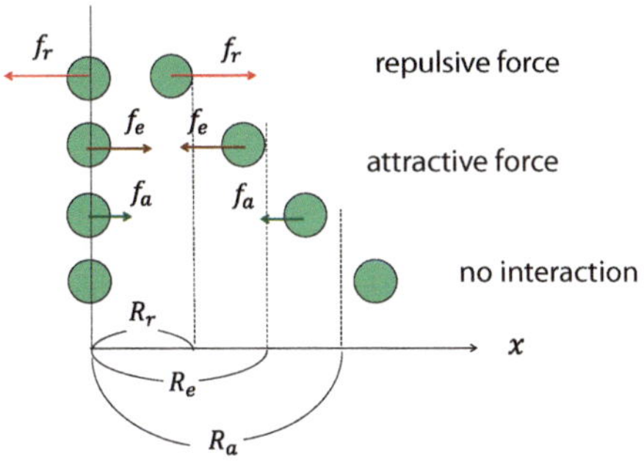

Fig. 2.10 Schematic figure of two-body interaction in the present model

cell nucleus R_r ($\sim$20 μm). Here excluded volume of an EC is the volume that is inaccessible to other ECs in a vessel as a result of presence of the first EC. Because of excluded volume, repulsive interaction arises between two ECs when they get closer. We denote by f_r the strength of the repulsive force. Since ECs frequently change their shape with extending or shrinking their pseudopodia, we denote by R_a (R_e) the length that represents the maximum (minimum) size of ECs. We estimate that R_e is 2 $\sim$ 3 times R_r and that R_a is 3 $\sim$ 5 times R_r. If the distance between two ECs is greater than R_a, no interaction takes place between them, while it is less than R_a, attractive interaction can be induced by cell contact. Strength of the attractive force between two ECs presumably depends on the distance between them and becomes stronger as the two ECs get closer as long as excluded volume effect does not have significant influence. Parameters f_e, f_a denote the strengths of the attractive interactions. Effects of these interactions are not instantaneous contrary to particle–particle interactions via physical fields such as electromagnetic fields or gravitational field, and there exists some specific response time in this system, which means that a simple Newtonian equation of motion defined by differential equations would not be suitable for modeling the system. If we consider x_n^t, v_n^t as the values of continuous functions $x_n(s)$, $v_n(s)$ at $s := t\Delta t$, noticing that we set Δt as a unit time, Eqs. (2.17) and (2.18) are rewritten as

$$x_n(s + \Delta t) = x_n(s) + \Delta t \cdot v_n(s) \tag{2.20}$$

$$v_n(s + \Delta t) - v_n(s) = \Delta t \cdot \left(-\gamma v_n(s) + \sum_{k \neq n} F(x_n(s) - x_k(s)) \right). \tag{2.21}$$

Equations (2.20) and (2.21) elucidate that time delay of interaction is incorporated into our model.

We do not consider the effect that ECs in the neogenetic vessels go back into the existing aortic ring, and x_n^t has to take nonnegative value. To avoid back flow of ECs into the source, we reset $x_n^{t+1} = 0$ and $v_n^{t+1} = 0$ if $x_n^{t+1} \leq 0$ in (2.17). Since we assume that the driving force of motility of ECs is sum of the two-body interactions, the velocity of an EC eventually happens to be quite high when the EC is surrounded by a large number of ECs, which is unrealistic because an EC moves mainly with its pseudopodia and the actual velocity of an EC is limited. Physically, it suggests that we have to take not only two-body interactions but multibody interactions into account. To incorporate this effect in a simple manner, we reset $|v_n^t|$ not to exceed a maximum value v_{max}. As for the parameters f_r, f_e, and f_a, we do not have any reasonable estimates; however, as we see below, the dynamics of ECs in our model does not qualitatively depend on the magnitude of interaction forces as far as we include both repulsive and attractive forces.

2.3.3 Cell-Mixing and Scaling Behavior of EC Dynamics

In order to see the effects of repulsive and attractive interactions, we simulated (2.17) and (2.18) for three types of two-body interactions: (a) only repulsive ($f_e \to -f_e$, $f_a \to -f_a$ in (2.19)), (b) only attractive ($f_r \to -f_r$), and (c) both interactions given in (2.19). A typical result is shown in Figs. 2.11 and 2.12, where we put $a_i = a$(constant) and $v_{ini}(i) = v_{ini}$(constant) for all i.

When interaction is only repulsive, the distribution of ECs is fairly uniform as shown in Fig. 2.12a, while if interaction is only attractive, ECs clump together as in Fig. 2.12b. In case both repulsive and attractive interactions coexist as is supposed in our model, ECs clearly show cell-mixing behavior and the distribution of ECs is sufficiently uniform. Cell-mixing behavior is also seen in Fig. 2.11a, though it is less frequent than Fig. 2.11c. The cell-mixing behavior depends on the time unit (Δt) in Eqs. (2.17) and (2.18). Figure 2.13a–c shows the trajectory of ECs with only repulsive interaction for $\Delta t = 0.5, 0.1, 0.05$. We find that cell-mixing behavior is suppressed for smaller Δt. The simulations in the case of both repulsive and

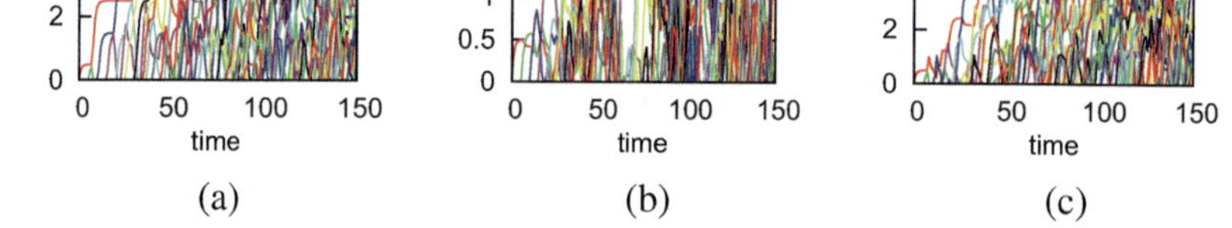

Fig. 2.11 Trajectories of EC movements during elongation with parameters $R_r = 0.2$, $R_e = 0.4$, $R_a = 0.7$, $\gamma = 0.6$, $a = 5$, $v_{ini} = 0.3$, $v_{max} = 1.0$, and (**a**) repulsive force only: $f_r = 0.6$, $f_e = 0$, $f_a = 0$, (**b**) attractive force only: $f_r = 0$, $f_e = 0.15$, $f_a = 0.05$, and (**c**) both: $f_r = 0.6$, $f_e = 0.15$, $f_a = 0.05$

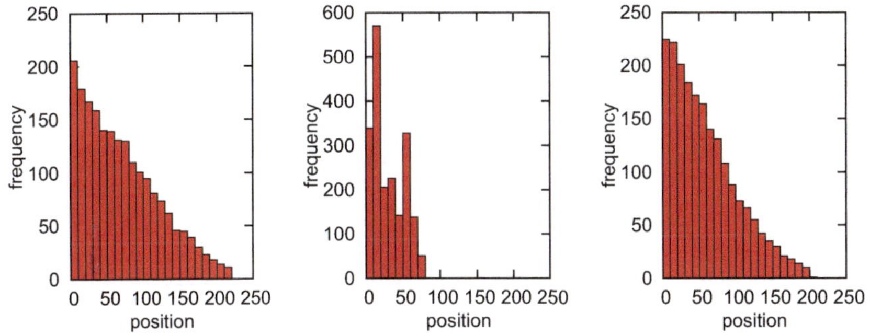

(a) Repulsive force only (b) Attractive force only (c) Both

Fig. 2.12 Distributions of ECs at time step $t = 10^4$. Parameters are the same as those in Fig. 2.11

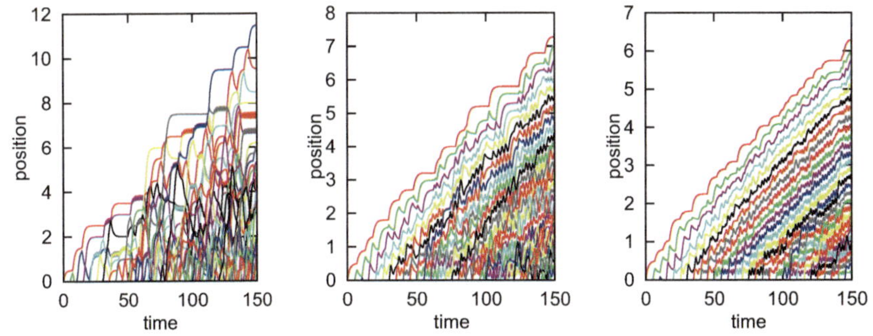

(a) (b) (c)

Fig. 2.13 Trajectories of EC movements during elongation for purely repulsive interactions with different time difference Δt. (**a**) $\Delta t = 0.5$, (**b**) $\Delta t = 0.1$, and (**c**) $\Delta t = 0.05$. The other parameters are the same as those in Fig. 2.11a

attractive interactions with the same parameters are shown in Fig. 2.14a–c. The cell-mixing behavior is a little suppressed at the tip, but it is still remarkable for small Δt. From these results, we conclude that repulsive interaction between ECs is necessary for smoothing the distribution of ECs and attractive interaction enhances cell-mixing behavior in the dynamics of ECs. The retardation of interaction also contributes to cell-mixing behavior.

We also show several simulation results for different parameters in Figs. 2.15, 2.16, and 2.17. Qualitative behavior does not change by different choice of the parameters. Figure 2.15a–c shows that the initial velocity of ECs has negligible impact on the elongation. This suggests that elongation is caused not by initial kinetic energy but by repulsive interactions among ECs. Figure 2.16b is the case that the ECs are injected with probability $1/5$ at every time step instead of the periodic injection $a = 5$. By comparing Figs. 2.15b and 2.16b, we may find that the

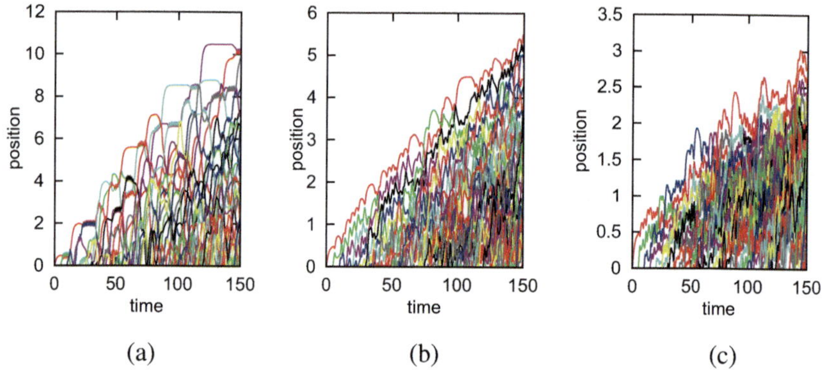

Fig. 2.14 Trajectories of EC movements during elongation for both repulsive and attractive interactions with different time difference Δt. (**a**) $\Delta t = 0.5$, (**b**) $\Delta t = 0.1$, and (**c**) $\Delta t = 0.05$. The other parameters are the same as those in Fig. 2.11c

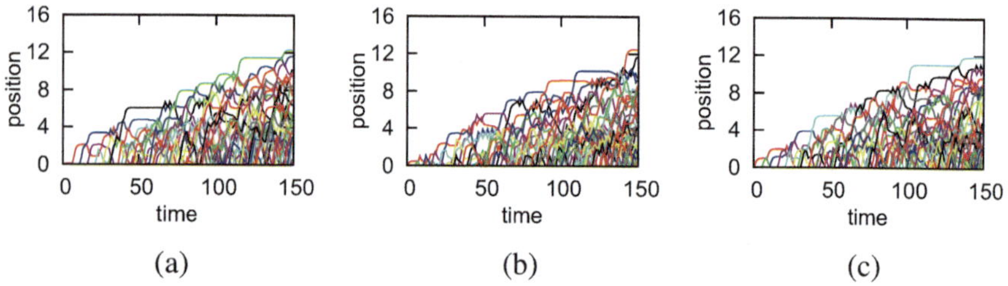

Fig. 2.15 Trajectories of ECs with the parameters that are the same as those in Fig. 2.11c except for v_{ini}. (**a**) $v_{ini} = 0.1$. (**b**) $v_{ini} = 0.3$. (**c**) $v_{ini} = 0.6$

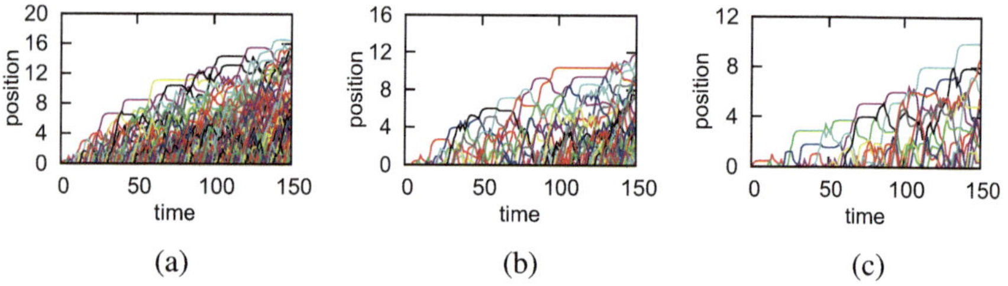

Fig. 2.16 Trajectories of ECs with the parameters that are the same as those in Fig. 2.11c except for the injection period a. In the center figure, "$\langle a \rangle = 5$" means that the injection period is random with average period 5. (**a**) $a = 2$. (**b**) $\langle a \rangle = 5$. (**c**) $a = 10$

way of injection of ECs scarcely affects the elongation as long as the mean interval $\langle a \rangle$ does not change. In fact, both periodic injection and random injection show the same scaling behavior of elongation length. This scaling behavior shall be discussed below (Figs. 2.18 and 2.19). Figures 2.16 and 2.17 show the robustness against the magnitude of interaction strength and the injection rate. These results support that

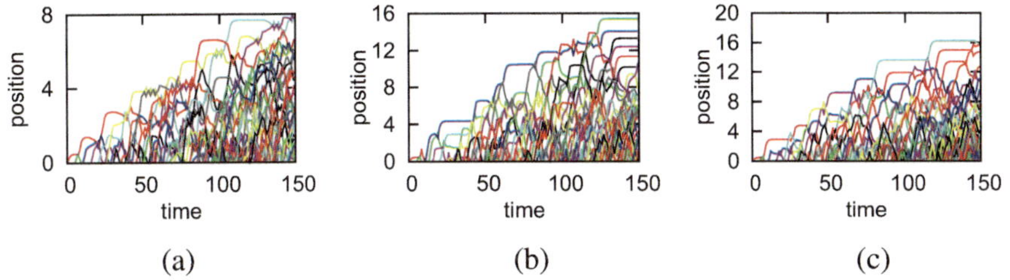

Fig. 2.17 Trajectories of ECs with the parameters that are the same as those in Fig. 2.11c except for the interaction strength (f_r, f_e, f_a). In the figures, "$\times m$" means that each of (f_r, f_e, f_a) is m times larger than that of Fig. 2.11c. (**a**) $\times 0.5$. (**b**) $\times 2$. (**c**) $\times 4$

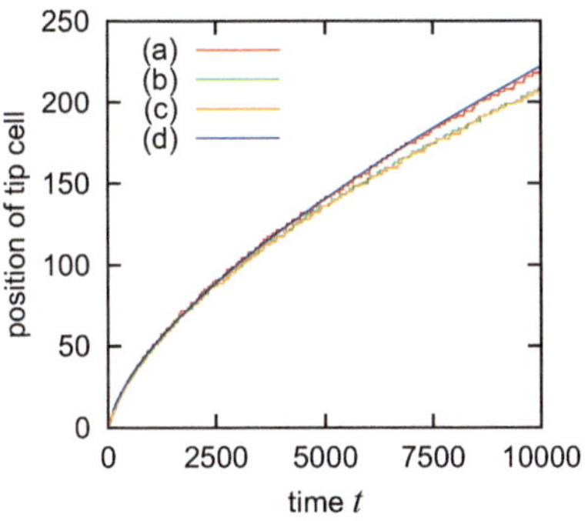

Fig. 2.18 Temporal change of position of the tip cell. The parameters are $R_r = 0.2$, $R_e = 0.4$, $R_a = 0.7$, $f_r = 0.6$, $f_a = 0$, $v_{ini} = 0$, $\gamma = 0.6$, $v_{max} = 1.0$, $\langle a \rangle = 5$, and (**a**) $f_e = 0$, (**b**) $f_e = 0.2$, and (**c**) $f_e = 0.4$. The curve (**d**) satisfies $y = 0.5092t^{0.6601}$

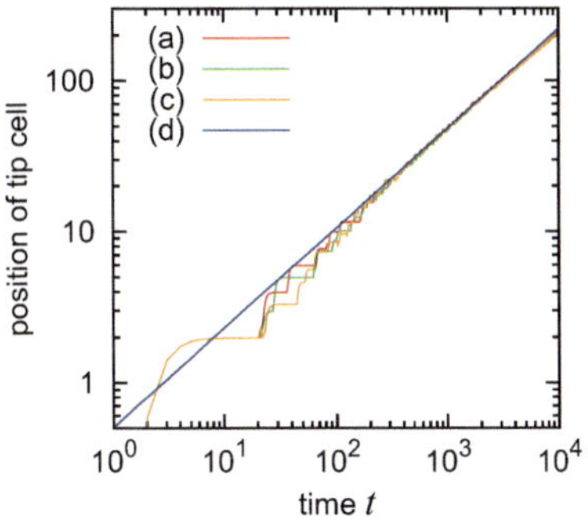

Fig. 2.19 Log–log plot of the curves (**a**)–(**d**) in Fig. 2.18. The straight line (**d**) satisfies $\ln(y) = 0.6601 \ln(t) - 0.6749$, which is fitted to the curve (**a**)

our model has potential for application to the cases of a wide range. Estimation of the parameters was recently performed with the experiments by Takubo et al.[60]

Figures 2.18 and 2.19 show the time dependence of the reaching position of ECs, that is, the position of an EC at the tip. As is seen from Fig. 2.19, it almost scales as $t^{2/3}$ for different strengths of attractive interaction. In fact we can find that this exponent 2/3 is observed almost irrespective of the parameters for interactions, injection rate, and initial velocity. In particular, the data of numerical simulation closely fit the curves $\propto t^{2/3}$ in case of no attractive interaction. For diffusive motion or random walk, the reach distance scales as $t^{1/2}$, while it scales as t^1 for ballistic motion or wave propagation. The exponent 2/3 suggests that the ECs in the present model show the dynamics between diffusive motion and ballistic motion. This universal value of the exponent could be understood as follows.

Let $\rho(x,t)$ be the density of the ECs at time t and position x. In continuum limit, it satisfies the equation of continuity:

$$\frac{\partial \rho(x,t)}{\partial t} + \frac{\partial}{\partial x}\left(\rho(x,t)v(x,t)\right) = \frac{1}{a(t)}\delta(x), \tag{2.22}$$

where $v(x,t)$ is the field of velocity of ECs at (x,t), and $a(t)$ is the injection rate of ECs. From (2.18), in quasi-equilibrium, we may approximate $v(x,t)$ as

$$v(x,t) \sim \frac{1}{\gamma}\int F(x-y)\rho(y,t)\,dy$$

If $\rho(x,t)$ is a smooth function of x, we can expand

$$\rho(y,t) = \rho(x,t) + \frac{\partial \rho(x,t)}{\partial x}(y-x) + \frac{\partial^2 \rho(x,t)}{\partial x^2}\frac{(y-x)^2}{2} + \cdots.$$

We further assume that higher delivatives of $\rho(x,t)$ take smaller absolute values than its first delivative $\frac{\partial \rho(x,t)}{\partial x}$. In fact, as shown in Fig. 2.3a and c, the density $\rho(x,t)$ is approximately a linear function, which implies that its higher order delivatives are negligible. Since $F(x)$ is an odd function, neglecting the 3rd-order terms and higher than it, we have

$$\begin{aligned} v(x,t) &\sim \frac{1}{\gamma}\int (y-x)F(x-y)\frac{\partial \rho}{\partial x}(x,t)\,dy \\ &=: A_{\mathrm{eff}}\frac{\partial \rho}{\partial x}(x,t), \end{aligned}$$

where A_{eff} is a constant depending on the parameters $\gamma,\ f_r,\ f_e,\ f_a,\ R_r,\ R_e,\ R_a$. Hence (2.22) gives

$$\frac{\partial \rho(x,t)}{\partial t} + A_{\mathrm{eff}}\frac{\partial}{\partial x}\left(\rho(x,t)\frac{\partial \rho}{\partial x}(x,t)\right) = \frac{1}{a(t)}\delta(x). \tag{2.23}$$

Suppose that $\rho(x,t)$ has a scaling form with exponents $\alpha,\ \beta$ as

$$\rho(x,t) = t^{\alpha} G(x/t^{\beta}). \tag{2.24}$$

Then, from (2.23), we have

$$t^{\alpha-1}\left\{\alpha G(X) - \beta X G'(X)\right\} + t^{2(\alpha-\beta)} A_{\text{eff}}\left[G'(X)^2 + G(X)G''(X)\right]$$
$$\sim \frac{1}{a(t)}\delta(t^{\beta}X) \quad (X := x/t^{\beta}).$$

Since

$$\delta(t^{\beta}X) = |t^{-\beta}|\delta(X),$$

and, if $a(t)$ is almost constant in average, we may replace $a(t)$ with its average $\langle a\rangle$, we have

$$t^{\alpha-1}\left\{\alpha G(X) - \beta X G'(X)\right\} + t^{2(\alpha-\beta)} A_{\text{eff}}\left[G'(X)^2 + G(X)G''(X)\right] \sim \frac{t^{-\beta}}{\langle a\rangle}\delta(X).$$

Hence, under the assumption of the scaling form (2.24), we find that

$$\alpha - 1 = 2(\alpha-\beta) = -\beta. \tag{2.25}$$

Thus, we find

$$\alpha = \frac{1}{3}, \quad \beta = \frac{2}{3}. \tag{2.26}$$

The length of the vessel, $l(t)$, is given the first zero of the function $G(X)$. If we denote by X_0 the position of this zero, we find $l(t)t^{-\beta} = X_0$ Therefore, $l(t) \propto t^{2/3}$, and the reaching position of ECs scales as $t^{2/3}$.

Next we consider the parameter dependence of cell-mixing behavior. Figure 2.20 shows the effect of attractive interaction f_e on cell mixing. Here N_{mix} is the number of cell mixing at the tip from $t = 0$ to $t = 150$. Its dependence on f_e is fairly complex, and so far we have not analyzed the details. However, numerical simulations show that the qualitative behavior of N_{mix} as a function of f_e is not sensitive to the parameter γ. The cell-mixing takes place even in the case of repulsive interactions ($f_e < 0$). A reason why cell-mixing takes place is due to the existence of oscillating bound states of two ECs that are observed in numerical

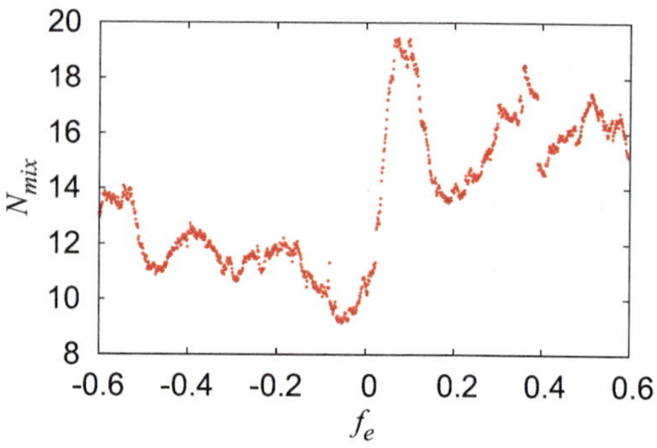

Fig. 2.20 Parameter f_e dependence of N_{mix} at the tip from $t = 0$ to $t = 150$, where N_{mix} denotes the number of times of change in leading EC. Each point shows the average number of 10^5 trials and $R_r = 0.2$, $R_e = 0.4$, $f_r = 0.6$, $f_a = 0$, $\gamma = 0.6$, $v_{ini} = 0.3$, and $\langle a \rangle = 5$

simulations. Let us consider two ECs whose positions at t_0 are $x_1^{t_0} = X_0 + \xi^{t_0}$, $x_2^{t_0} = X_0 - \xi^{t_0}$ with velocity $v_1^{t_0} = v^{t_0}$ and $v_2^{t_0} = -v^{t_0}$. If there is no interaction from the other ECs, it holds that

$$\xi^t := x_1^t - X_0 = -(x_2^t - X_0), \quad v^t := v_1^t = -v_2^t$$

for $t \geq t_0$ because of the symmetry of interaction. Furthermore we assume the repulsive interactions $f_r > 0$, $f_e = f_a = 0$ for simplicity, then (2.17) and (2.18) yield the following:

$$\xi^{t+1} = \xi^t + v^t \tag{2.27}$$

$$v^{t+1} = (1-\gamma)v^t + \begin{cases} f_r & 0 < \xi^t < \frac{1}{2}R_r \\ -f_r & -\frac{1}{2}R_r < \xi^t < 0 \\ 0 & \text{otherwise} \end{cases}. \tag{2.28}$$

Suppose that

$$v^{t_0} = -\frac{(1-\gamma)f_r}{\gamma^2 - 3\gamma + 3}, \quad 0 < \xi^{t_0} < \frac{R_r}{2},$$

$$-\left(\xi^{t_0} + \frac{R_r}{2}\right) < v^{t_0} < \min\left[-\xi^{t_0}, (1-\gamma)\left(\xi^{t_0} - \frac{R_r}{2}\right)\right].$$

Then, we have a periodic solution to (2.27) and (2.28)

$$(\xi^{t_0}, v^{t_0}) = (\xi^{t_0+3}, v^{t_0+3}).$$

For example, let $\gamma = \frac{3}{5}$, $R_r = \frac{1}{5}$, $f_r = \frac{3}{5}$, $\xi^{t_0} = \frac{1}{13}$, $v^{t_0} = -\frac{2}{13}$, which satisfy the above conditions. Then, we have a sequence (ξ^t, v^t) as

$$\left(\frac{1}{13}, -\frac{2}{13}\right) \to \left(-\frac{1}{13}, \frac{7}{13}\right) \to \left(\frac{6}{13}, -\frac{5}{13}\right) \to \left(\frac{1}{13}, -\frac{2}{13}\right) \to \cdots,$$

which shows period 3. The oscillating bound states of this kind give rise to the cell-mixing behavior of ECs even if interactions are merely repulsive. Note that there is no such oscillating bound state in continuous limit of (2.27) and (2.28):

$$\frac{d\xi}{dt}(t) = v(t)$$
$$\frac{dv}{dt}(t) = -\gamma v(t) + F(2\xi(t)).$$

The only stationary state for two ECs is a bound state with velocity 0. An oscillating bound state appears because of discretization, which intrinsically reflects the retardation of reaction in ECs.

2.3.4 Pattern Formation of the Model with Elongation and Bifurcation of Neogenetic Vessels

Now let us consider the process of elongation and bifurcation of vessels. Important observations in the experiments are (a) a vessel divides into not more than two branches at a junction, (b) bifurcation occurs mostly at the tip when the tip is crowded by ECs, and (c) a vessel splits into two branches with an angle of approximately 60 degrees[4]. As is explained in the previous section, the blood vessel network we consider is the early stage of angiogenesis and is modeled by a tree graph. The first neogenetic vessel is referred to the branch with index 1, and the branches appeared by bifurcations are indexed by positive integers in order. The index of the branch in which nth EC exists at $t \in \mathbb{Z}_{\geq 0}$ is denoted by $\mu(n,t) \in \mathbb{Z}_+$, and the position of the EC is denoted by the same notation x_n^t (≥ 0) that, however, is measured from the origin of the branch $\mu(n,t)$. The position of the tip of the branch ν is denoted by b_ν^t (> 0). In view of the observations (a) and (b), we introduce three new parameters X_e, X_b (> 0), and $L_b \in \mathbb{Z}_{>0}$, which indicate the threshold impulse for elongation, that for bifurcation, and threshold congestion number of ECs at the tip, respectively. When we closely observe the process of bifurcation in experiments, we notice that an EC remains at the tip and other two ECs pass through it and that one slips through one side of the impassive EC and the other through the opposite side. Hence we impose the condition $L_b = 3$. The time evolution of our model is determined as follows.

While the first bifurcation has not taken place (i.e.,$^\forall n,\ \mu(n,t) = 1$), we put

$$\tilde{x}_n^{t+1} := x_n^t + v_n^t,$$
$$\tilde{v}_n^{t+1} := v_n^t - \gamma v_n^t + \sum_{k \neq n} F(x_n^t - x_k^t),$$

$$X^t := \sum_{k=1}^{N(t)} \max[0, \tilde{x}_k^{t+1} - b_1^t],$$

$$L^t := \sum_{k=1}^{N(t)} \theta(\tilde{x}_k^{t+1} - b_1^t).$$

Here $\tilde{x}_n^t$ ($\tilde{v}_n^t$) denotes the provisional position (velocity) of nth EC at time step t, X^t the impulse at the tip, L^t the congestion number at the tip, and θ the step function:

$$\theta(x) := \begin{cases} 1 & (x > 0) \\ 0 & (x \leq 0) \end{cases}.$$

Note that we set $\tilde{v}_n^{t+1} \to \pm v_{max}$ when $\left|\tilde{v}_n^{t+1}\right| > v_{max}$ as in Sect. 2.3.2. Then the updating rule is given as (Fig. 2.21):

(i) if $X^t < X_e$, $x_n^{t+1} = \min[\tilde{x}_n^{t+1}, b_1^t]$, $v_n^{t+1} = \tilde{v}_n^{t+1}$ for all n and $b_1^{t+1} = b_1^t$.
(ii) Else if $X^t < X_b$ or $L^t < L_b$, then $b_1^{t+1} = b_1^t + \Delta b$, and $x_n^{t+1} = \min[\tilde{x}_n^{t+1}, b_1^{t+1}]$, $v_n^{t+1} = \tilde{v}_n^{t+1}$ for all n. Here Δb is the positive parameter that denotes elongation velocity at the tip of a vessel.
(iii) Else, that is, $X_b \leq X^t$ and $L_b \leq L^t$, then the vessel bifurcates into two branches. We put $b_1^{t+1} = b_1^t$, $b_2^{t+1} = b_1^t + \Delta b$, and $b_3^{t+1} = b_1^t + \Delta b$. Hence $b_1^{t+s} = b_1^t$ for $s \geq 1$. Let $\mathcal{K} := \{k_1, k_2, \ldots, k_m\}$ be the set of indices of ECs that satisfy $\tilde{x}_{k_i}^{t+1} > b_1^t$. We assume $\tilde{x}_{k_1}^{t+1} \leq \tilde{x}_{k_2}^{t+1} \leq \ldots \leq \tilde{x}_{k_m}^{t+1}$. Then, if $n \notin \mathcal{K}$, $x_n^{t+1} = \tilde{x}_n^{t+1}$, $v_n^{t+1} = \tilde{v}_n^{t+1}$, and $\mu(n, t+1) = 1$. For $n \in \mathcal{K}$, $x_{k_1}^{t+1} = b_1^t$, $v_{k_1}^{t+1} = 0$, $\mu(k_1, t+1) = 1$, $x_{k_{2i}}^{t+1} = b_1^t + \Delta b$, $v_{k_{2i}}^{t+1} = \tilde{v}_{k_{2i}}^{t+1}/2$, $\mu(k_{2i}, t+1) = 2$, $x_{k_{2i+1}}^{t+1} = b_1^t + \Delta b$, $v_{k_{2i+1}}^{t+1} = \tilde{v}_{k_{2i}}^{t+1}/2$, and $\mu(k_{2i+1}, t+1) = 3$ $(i = 1, 2, \ldots)$.

After bifurcations took place, we apply the above updating rule to the ECs in each branch. Two ECs in separate branches can interact with each other only if one branch has bifurcated from the other branch. At a junction of bifurcation, we assume that an EC moves into one branch from which attractive force is stronger than that from the other branch. In case of the equal strength of attractive force, the EC selects one of them with equal probability.

Apparently the condition $X_e < X_b$ must be satisfied and the order of the magnitude of X_e and X_b is supposed to be of the same order of that of the repulsive force f_r. The parameter X_e is also dependent on the stiffness of the collagen gel in which ECs are cultivated or that of extracellular organic materials' in vivo experiments. The injection rate a depends on the size of the supply source (existing blood vessels). It is also affected by substances to control cell proliferation such as VEGF that also affects the parameters X_e, X_b by activation of pseudopodia. The parameter Δb gives the scale of an overall vessel network and does not change the pattern of its shape. Hence we set $\Delta b = 1$ in numerical simulations. The

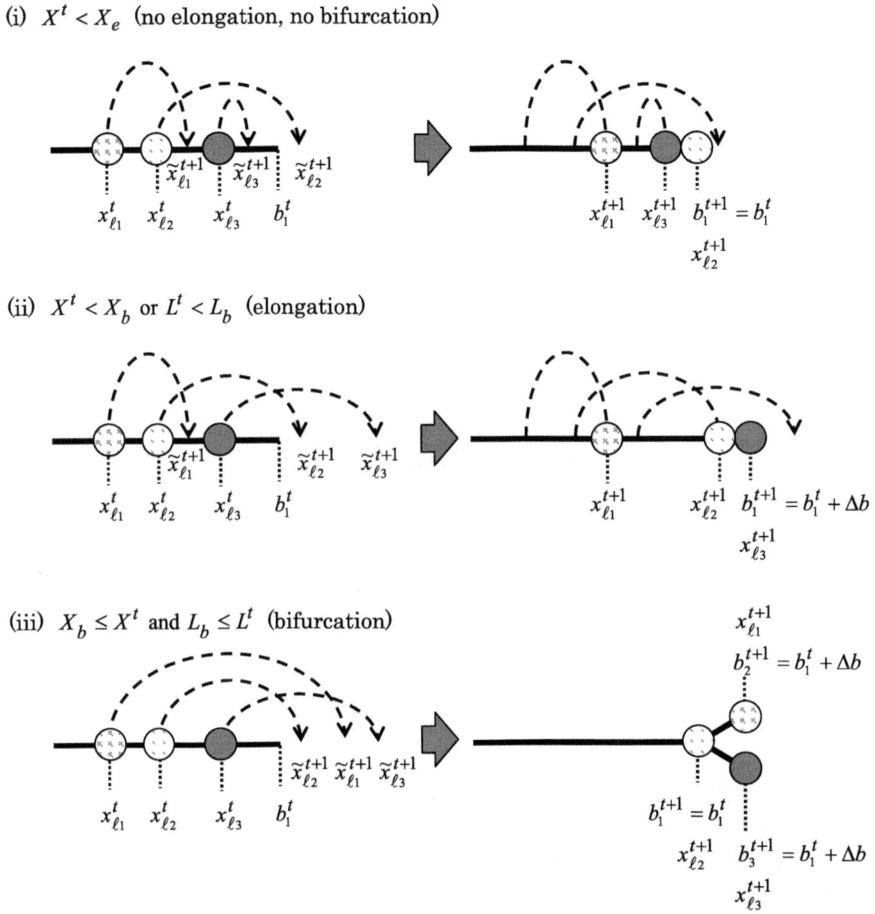

Fig. 2.21 Schematic representation of the updating rule for elongation and bifurcation. (i) If $X^t < X_e$, neither elongation nor bifurcation takes place. (ii) Else if $X_t \le X_b$ or $L^t < L_b$, the vessel grows by Δb. (iii) Else, that is, $X_b \le X^t$ and $L_b \le L^t$, the tip bifurcates into two branches. According to the rule (iii) in the text, $\mathscr{K} = \{\ell_1, \ell_2, \ell_3\}$, $k_1 = \ell_2$, $k_2 = \ell_1$, $k_3 = \ell_3$, and $x_{k_1}^{t+1} = b_1^{t+1} = b_1^t$, $x_{k_2}^{t+1} = b_2^{t+1} = b_1^t + \Delta b$, $x_{k_3}^{t+1} = b_3^{t+1} = b_1^t + \Delta b$

typical growth patterns of blood vessels in the present model are shown in Figs. 2.22 and 2.23, where small red discs denote the positions of nuclei of ECs, white tubes denote the neogenetic vessels, and we take the observation (c) into account at the junctions. Since we have not considered anastomosis, these figures represent the projection of vessel networks onto a two-dimensional plane. Though some vessels look like recombined, they are apart in the direction perpendicular to the plane. We also show an example of retina-like vessel patterns in Fig. 2.24 in which neogenetic vessels elongate in four directions from the origin.

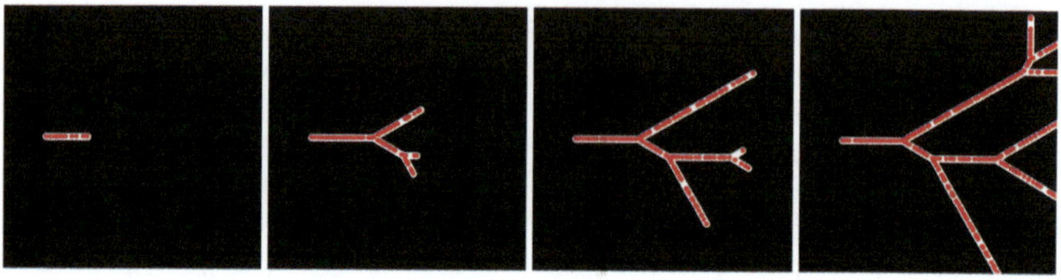

Fig. 2.22 A simulation of angiogenesis using our model with $X_e = 1.4$, $X_b = 2.0$, $L_b = 3$, $f_r = 0.6$, $f_e = 0.15$, $f_a = 0.05$, $R_r = 0.2$, $R_e = 0.4$, $R_a = 0.7$, $\gamma = 0.6$, $a = 5$, $v_{ini} = 0.3$, and $v_{max} = 1$. Red dots denote the positions of nuclei of ECs. The first vessel is placed on the left edge of the white tube as the initial condition. The patterns are snapshots at time steps 100, 500, 1000, and 2000, respectively, from left to right

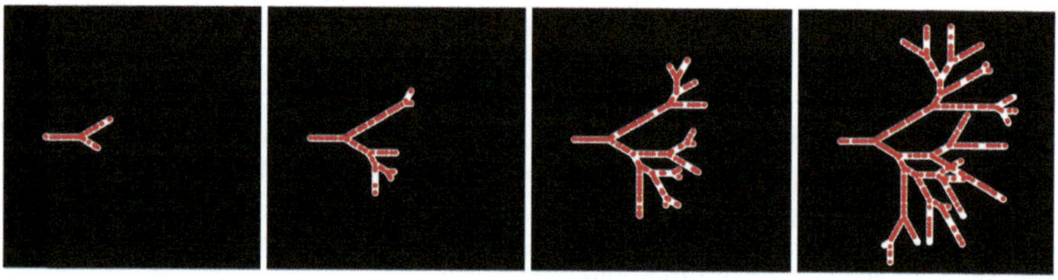

Fig. 2.23 A simulation with different parameters, where $X_e = 1.4$, $X_b = 1.6$, $L_b = 3$, $f_r = 0.6$, $f_e = 0.15$, $f_a = 0.05$, $R_r = 0.2$, $R_e = 0.4$, $R_a = 0.7$, $\gamma = 0.6$, $a = 5$, $v_{ini} = 0.3$, and $v_{max} = 1$. Time steps are 200, 500, 1000, and 2000, respectively, from left

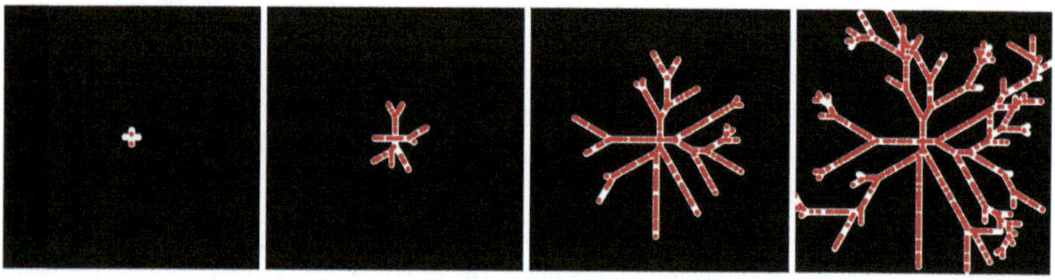

Fig. 2.24 A simulation of angiogenesis from the center with four initial vessels: $X_e = 1.9$, $X_b = 2.2$, $L_b = 3$, $f_r = 0.6$, $f_e = 0.15$, $f_a = 0.05$, $R_r = 0.2$, $R_e = 0.4$, $R_a = 0.7$, $\gamma = 0.6$, $a = 5$, $v_{ini} = 0.3$, and $v_{max} = 1$. The patterns are the snapshots at time step $t = 10$, 400, 1600, and 4000, respectively, from left

Finally we show the difference of sprouting patterns by changing X_e and keeping all else fixed in Fig. 2.25. We suppose the difference of X_e corresponds to that of stiffness of collagen. As X_e approaches X_b, the average length of branches becomes shorter and fluctuation of lengths becomes larger.

For a given injection sequence $\{a_i\}$ $(i \in \mathbb{Z}_+)$, the pattern of a simulated vessel network is determined almost uniquely because the time evolution rules of the present model are deterministic except when an EC near a junction receives the same strength of attractive interaction from the two branches. However, a time evolution

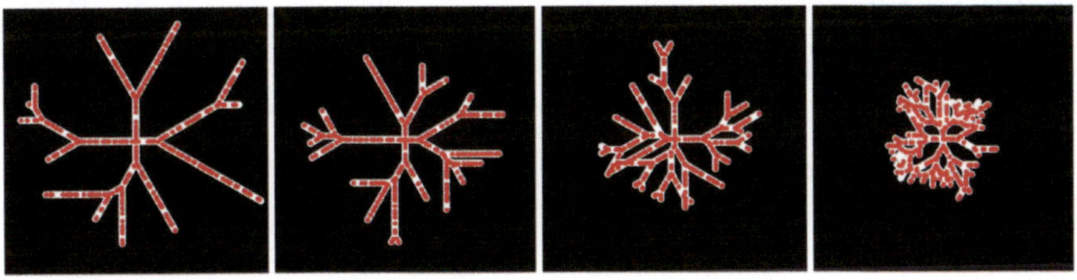

Fig. 2.25 The sprouting patterns at time step 2000 for $X_e = 1.8, 1.9, 2.0$, and 2.1, respectively, from left. $X_b = 2.3$ and other parameters are the same as those in Fig. 2.24. Total number of ECs in these patterns is the same (400)

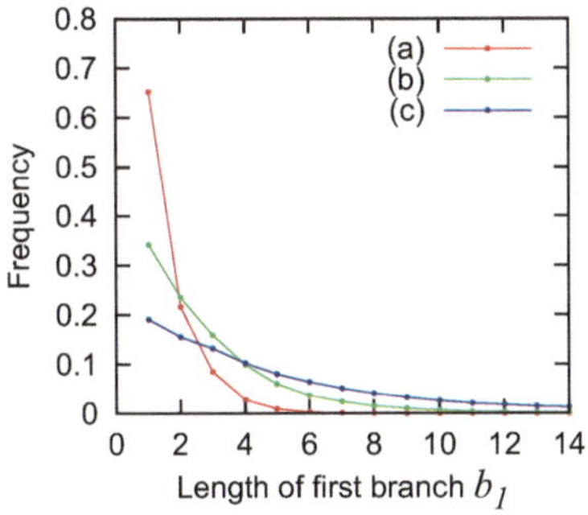

Fig. 2.26 The distributions of the lengths of the first branches. The threshold values are $X_e = 2.0$, **(a)** $X_b = 2.1$, **(b)** $X_b = 2.3$, and **(c)** $X_b = 2.5$, and the other parameters are the same as those in Fig. 2.22

pattern can easily change by slight modification of injection interval. If ECs are injected randomly with an average rate $\langle a \rangle^{-1}$, a pattern of neogenetic blood vessels differs at each trial of simulations. The patterns appear to be qualitatively similar, but there is little reproducibility in the patterns. Let b_1 be the length of the first branch, i.e., the length between the origin of the blood vessel and the first junction. This length of first branch does not change any more after the first bifurcation due to the above updating rule (iii) as $b_1^{t+s} = b_1^t$ for all $s \geq 1$. The distributions of the length of first branch b_1 are shown in Figs. 2.26 and 2.27.

We find that the distribution $f(b_1)$ $(0 \leq f(b_1) < 1$ and $\int_0^\infty f(b_1)\, db_1 = 1$) of the length of first branch b_1 clearly shows power-law behavior:

$$f(b_1) \propto \frac{1}{K^{b_1}},$$

where K $(K > 1)$ is a constant determined by the parameters. From Figs. 2.26 and 2.27, we find that K is a decreasing function of X_b.

Although the length of branch has wide dispersion, the dependence of the average values on the parameters X_e and X_b is fairly clear. It becomes shorter as the value

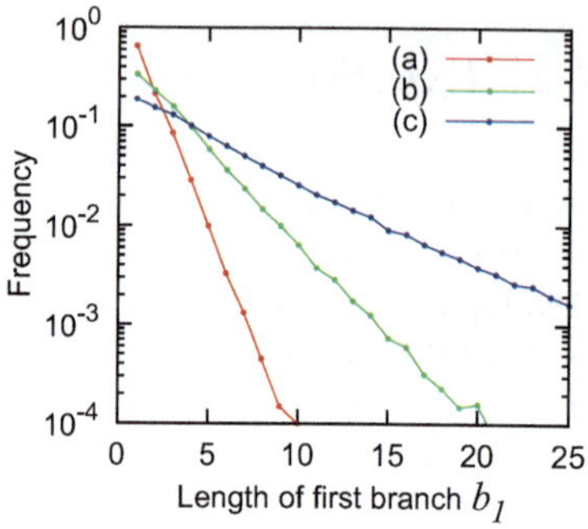

Fig. 2.27 Log plot of the curves **(a)**–**(c)** in Fig. 2.26. The distributions show clear power-law behavior

Table 2.1 The average length of the first blood vessel is shown for various parameters X_e and X_b. The other parameters are fixed as $L_b = 3$, $f_r = 0.6$, $f_e = 0.15$, $f_a = 0.05$, $R_r = 0.2$, $R_e = 0.4$, $R_a = 0.7$, $\gamma = 0.6$, $\langle a \rangle = 5$, and $v_{ini} = 0.3$. Number of trials for each parameter is 10^5. The N/A stands for the parameter $X_b \leq X_e$ that does not satisfy the requirement, and "-" for rare bifurcation because of the severe condition on bifurcation $X_b \gg X_e$

$X_b \backslash X_e$	1.1	1.2	1.3	1.4	1.5	1.6	1.7	1.8	1.9	2.0
1.2	4.66	N/A	N/A	N/A	N/A	N/A	N/A	N/A	N/A	N/A
1.3	6.03	3.71	N/A	N/A	N/A	N/A	N/A	N/A	N/A	N/A
1.4	8.28	4.76	3.11	N/A	N/A	N/A	N/A	N/A	N/A	N/A
1.5	12.7	6.36	3.94	2.66	N/A	N/A	N/A	N/A	N/A	N/A
1.6	–	9.12	5.19	3.37	2.33	N/A	N/A	N/A	N/A	N/A
1.7	–	14.6	7.15	4.38	2.91	2.04	N/A	N/A	N/A	N/A
1.8	–	–	10.9	6.01	3.80	2.58	1.85	N/A	N/A	N/A
1.9	–	–	18.7	8.91	5.26	3.38	2.38	1.70	N/A	N/A
2.0	–	–	–	14.6	7.69	4.64	3.10	2.17	1.56	N/A
2.1	–	–	–	–	12.7	6.96	4.41	2.94	2.07	1.54
2.2	–	–	–	–	25.8	11.6	6.57	4.14	2.84	2.05
2.3	–	–	–	–	–	21.7	10.4	6.08	3.99	2.81
2.4	–	–	–	–	–	–	18.5	9.39	5.66	3.85
2.5	–	–	–	–	–	–	–	15.6	8.45	5.36
2.6	–	–	–	–	–	–	–	–	13.6	7.92
2.7	–	–	–	–	–	–	–	–	–	12.8
2.8	–	–	–	–	–	–	–	–	–	23.5

X_e increases, while it becomes longer while the value X_b increases. In Table 2.1, we show dependence of the length of first blood vessel on the parameters X_e and X_b. These values are the mean values of 10^5 trials for each set of parameter values. The larger the threshold for bifurcation X_b becomes, the longer the length of blood vessel tends to be. For the elongation threshold X_e, the inverse dependence is observed.

2.3.5 A Continuous Model for Angiogenesis

In this subsection, we give a mathematical model for elongation and bifurcation of blood vessels described by simultaneous ordinary differential equations[38]. We pay attention to one neogenetic blood vessel that will elongate and bifurcate to construct a network of blood vessels. Let the initial length of the blood vessel be L_{ini} at initial time t_0. Noticing the fact that a blood vessel bifurcates into only two and that the number of branches arose after kth order bifurcation is 2^k, we denote by the (k, i) branch the ith branch ($i = 1, 2, \ldots, 2^k$) produced after kth order bifurcation of the blood vessel. The simultaneous ordinary differential equations we treat here are given as

$$\frac{d}{dt}L_{k+1}^{(i)}(t) = F(n_{k+1}^{(i)}(t)) \quad (0 \le n_{k+1}^{(i)}(t) < n_b,\ k = 0, 1, 2, \ldots,\ i = 1, 2, \ldots, 2^k), \tag{2.29}$$

where $L_{k+1}^{(i)}(t)$ denotes the length of the (k, i) branch, and we assume that it bifurcates when the number of ECs at the tip, $n_{k+1}^{(i)}(t)$, exceeds the threshold value n_b. The function $F(n)$, which will be defined afterward, describes the speed of elongation, $n_{k+1}^{(i)}(t)$, and the density of ECs, $\rho(t)$, are determined by the total number of ECs, $N(t)$, as follows:

$$\rho(t) = \frac{N(t)}{s_1 L_1(t) + \sum_{k=1}^{\infty} \sum_{i=1}^{2^k} s_{k+1}^{(i)} L_{k+1}^{(i)}(t)}, \tag{2.30}$$

$$n_{k+1}^{(i)}(t) = \lambda_{k+1}^{(i)}(t) \left(\rho(t) s_{k+1}^{(i)} l \right). \tag{2.31}$$

Here $s_k^{(i)}$ is the effective cross section of the (k, i) branch, l a parameter of the tip length, and $\lambda_k^{(i)}(t)$ that of the activation of ECs due to the VEGF concentration. We summarize the variables and parameters in Table 2.2. Although the system looks very complicated due to a lot of variables and parameters, once the parameters $s_k^{(i)}$, l and a function $F(n)$, which characterize the system, are determined and functions $N(t)$ and $\lambda_k^{(i)}(t)$, which are estimated from experimental settings, are given, then we can obtain a closed simultaneous equations of $L_k^{(i)}(t)$ through Eqs. (2.30) and (2.31). We will explain below the meaning of these parameters and equations step by step.

We supposed that there is a neogenetic blood vessel with length L_{ini} at $t = t_0$. Though it does not have lumen structure, we assume that we can define a cross section s_1 that denotes effective occupation area of ECs in the vessel. Since only the ECs near the tip will contribute to elongation and bifurcation of the vessel, we assume that the ECs that stay at a distance not more than l from the tip can affect elongation and bifurcation. Let $N(t)$ be the total number of ECs in the vessel, $L_1(t)$

Table 2.2 List of variables

Variable	Definition
$L_{k+1}^{(i)}(t)$	Length of ith neogenetic blood vessel of kth bifurcation at t $(1 \leq i \leq 2^k)$
$N(t)$	Total number of endothelial cells in neogenetic blood vessels at t
$\rho(t)$	Average density of endothelial cells in neogenetic blood vessels at t
$s_{k+1}^{(i)}$	Effective cross section of ith neogenetic vessel of kth bifurcation
$n_{k+1}^{(i)}(t)$	Effective number of endothelial cells at the tip of ith neogenetic vessel of kth bifurcation
$\lambda_{k+1}^{(i)}(t)$	VEGF activity at the tip of ith neogenetic blood vessel of kth bifurcation at t
l	Effective length of a tip relating to elongation and bifurcation
n_e	Threshold value of number of endothelial cells for elongation
n_b	Threshold value of number of endothelial cells for bifurcation
L_{ini}	Length of the initial blood vessel at initial $t = t_0$
N_{ini}	Number of endothelial cells at initial time $t = t_0$
$L_1(t)$	Length of the first blood vessel at t
s_1	Cross section of the first blood vessel at t
$\tilde{n}_1(t)$	Number of endothelial cells in the tip of the first blood vessel at t
$n_1(t)$	Effective number of endothelial cells in the tip of the first blood vessel at t
$t_k^{(i)}$	Bifurcation time of ith neogenetic blood vessel of kth bifurcation

the length of the vessel, and $\tilde{n}_1(t)$ the number of ECs at the tip. If the density of the ECs is almost constant in a vessel, we have

$$\tilde{n}_1(t) = \frac{N(t)l}{L_1(t)}. \tag{2.32}$$

The activation parameter at the tip due to VEGF, etc., is denoted by $\lambda_1(t)$, and we define $n_1(t) := \lambda_1(t)\tilde{n}_1(t)$. We assume that elongation and bifurcation are determined only by the variable $n_1(t)$; that is, there are two thresholds n_e, n_b $(n_e < n_b)$ such that elongation takes place when $n_e < n_1(t) \leq n_b$ and bifurcation takes place when $n_b < n_1(t)$. Then there exists a monotonically increasing function $F(n)$ that satisfies

$$\frac{d}{dt}L_1(t) = F(n_1(t)) \qquad (0 \leq n_1(t) < n_b). \tag{2.33}$$

The concrete form of $F(n)$ will be discussed later.

When bifurcation takes place at $t = t_1$, we have

$$n_1(t_1) = n_b, \tag{2.34}$$

and two new branches sprout. Let the cross sections of these new branches be $s_2^{(1)}$, $s_2^{(2)}$. Then, according to Murray's law, we have

$$(s_1)^m = (s_2^{(1)})^m + (s_2^{(2)})^m \qquad (1.4 \lesssim m \lesssim 1.6).$$

Since Murray's law is deduced from hydrodynamical consideration and will hold for complete blood vessels, it is not obvious that it is also true for neogenetic blood vessels under consideration; however, we assume that this law also holds in the later numerical simulations. Note that we use only the condition $s_1 > s_2^{(1)}, s_2^{(2)}$, and precise value m is not essential in the later discussions about exact solutions. We denote the length of the new branches by $L_2^{(1)}(t)$, $L_2^{(2)}(t)$, and the average density of ECs, $\rho(t)$, is given as

$$\rho(t) = \frac{N(t)}{s_1 L_1(t) + \sum_{i=1}^{2} s_2^{(i)} L_2^{(i)}(t)}. \tag{2.35}$$

Here $L_1(t) = L_1(t_1)$ $(t \geq t_1)$. Putting the activation parameter at each tips $\lambda_2^{(1)}$, $\lambda_2^{(2)}$, the effective number of ECs at each tip, $n_2^{(1)}(t)$, $n_2^{(2)}(t)$, is given as

$$n_2^{(i)}(t) = \lambda_2^{(i)}(t) \left(\rho(t) s_2^{(i)} l\right), \tag{2.36}$$

and similar equations to (2.33):

$$\frac{d}{dt} L_2^{(i)}(t) = F(n_2^{(i)}(t)) \qquad (0 \leq n_2^{(i)}(t) < n_b,\ i = 1, 2) \tag{2.37}$$

will hold and bifurcation takes place if $n_2^{(i)}(t) \geq n_b$. After bifurcation, the value of $L_2^{(i)}(t)$ remains constant. Similarly kth order bifurcation occurs 2^{k-1} times, and 2^k new branches are produced. Let the times at which these bifurcations occur be

$$t_k^{(1)} \leq t_k^{(2)} \leq \cdots \leq t_k^{(2^{k-1})},$$

effective cross sections $s_{k+1}^{(i)}$, the length of these branches at time t, $L_{k+1}^{(i)}(t)$, activation parameter $\lambda_{k+1}^{(i)}(t)$, and effective number of ECs at the tip $n_{k+1}^{(i)}(t)$ $(i = 1, 2, \ldots, 2^k)$. The density of ECs is given by the equation (2.30):

$$\rho(t) = \frac{N(t)}{s_1 L_1(t) + \sum_{k=1}^{\infty} \sum_{i=1}^{2^k} s_{k+1}^{(i)} L_{k+1}^{(i)}(t)},$$

where $L_{k+1}^{(j)} = 0$ ($^\forall j \geq 2i-1$) for $t < t_k^{(i)}$, and the length of a branch remains constant after bifurcation. Thus we have Eq. (2.31):

$$n_{k+1}^{(i)}(t) = \lambda_{k+1}^{(i)}(t)\left(\rho(t)s_{k+1}^{(i)}l\right),$$

and (2.29):

$$\frac{d}{dt}L_{k+1}^{(i)}(t) = F(n_{k+1}^{(i)}(t)) \qquad (0 \leq n_{k+1}^{(i)}(t) < n_b,\ i = 1, 2, \ldots, 2^k).$$

In summary, the present continuous model describes time evolution of the elongation and bifurcation of blood vessels by the time dependence of the branch lengths $L_k^{(i)}(t)$ calculated from differential equations (2.30), (2.31), and (2.29). The parameters, which may be estimated from experimental results, are the total number of ECs $N(t)$, activity due to VEGF etc. $\lambda_k^{(i)}(t)$, cross sections $s_k^{(i)}$, threshold values of elongation and bifurcation n_e, n_b, elongation function $F(n)$, initial length of the blood vessel L_{ini}, and the effective tip length l (Fig. 2.28).

Finally let us examine the function $N(t)$. In the time span of the experiment, cell division is rarely observed. Hence we may express $N(t)$ with a positive constant a as

$$N(t) = N_{ini} + a(t - t_0). \qquad (2.38)$$

Then, from (2.30) and (2.29),

$$\sum_k \sum_i s_k^{(i)} \frac{d}{dt} L_k^{(i)}(t) = \frac{d}{dt}\frac{N(t)}{\rho(t)} = \sum_k{}' \sum_i{}' s_k^{(i)} F(n_k^{(i)}(t)),$$

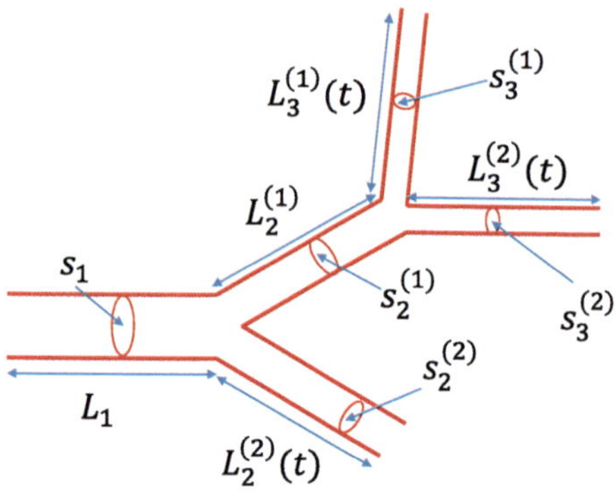

Fig. 2.28 Length of the branch $L_k^{(i)}$ and cross section $s_k^{(i)}$ in sprouting neogenetic blood vessel

where $\sum_k{}' \sum_i{}'$ denotes the summation over the growing branches, and we find

$$N(t)\frac{d}{dt}\rho(t) = \left\{a - \sum_k{}' \sum_i{}' s_k^{(i)} \rho(t) F\left(n_k^{(i)}(t)\right)\right\} \rho(t). \tag{2.39}$$

Therefore, if time dependence of the activity $\lambda_k^{(i)}(t)$ is negligible, i.e., $\lambda_k^{(i)}(t) = \lambda_k^{(i)}$, the density of ECs asymptotically approaches to the equilibrium value ρ_∞ determined by the following equation:

$$\sum_k{}' \sum_i{}' s_k^{(i)} \rho_\infty F\left(l\lambda_k^{(i)} s_k^{(i)} \rho_\infty\right) = a. \tag{2.40}$$

Equation (2.40) implies that the blood vessel bifurcates only finite times, but, from (2.38), ECs are injected into the network at the same rate. Consequently the last branches continue to extend unrealistically (Fig. 2.29). For a long span of time evolution, the effects of cell division of ECs in the blood vessel network are not negligible, but cell division will end after a certain period and accordingly supply of

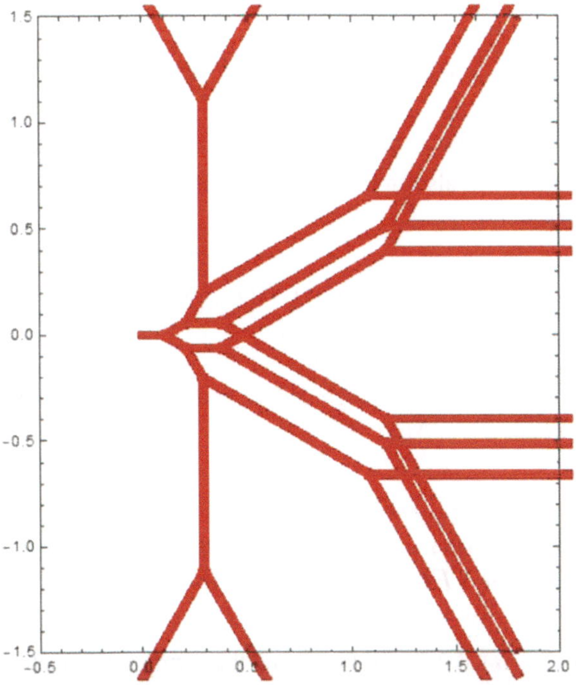

Fig. 2.29 Bifurcation of the present model when $N(t)$ satisfies (2.38). Here the parameters are $a = 50$, $n_e = 1$, $n_b = 3$, $s_1 = 3$, $\lambda_k^{(i)} = 1$, and we assume the Murray's law ($n = 1.5$)

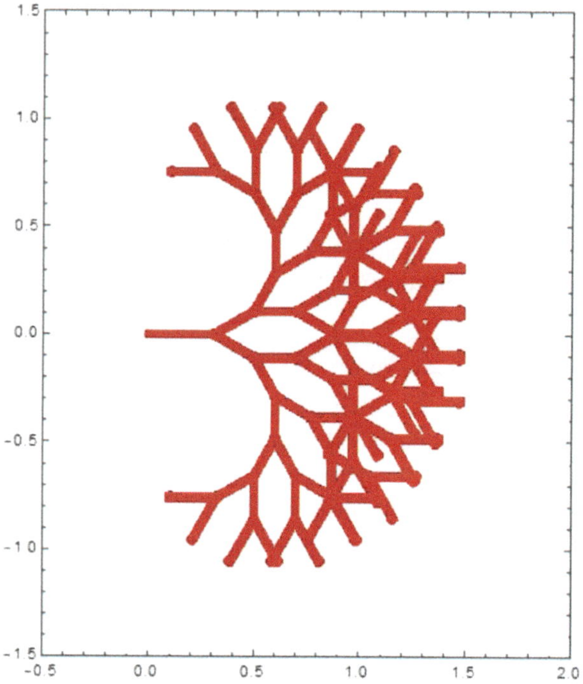

Fig. 2.30 Bifurcation of the present model when $N(t)$ is given by (2.41). The parameters are $\epsilon = 1.05$, $b = 1.05 \times 10^{-4}$, $N_{ini} = 10^{-3}$, and we assume the Murray's law ($n = 1.5$)

ECs from the origin will also stop. Hence $N(t)$ may satisfy the following Logistic equation (see Fig. 2.30):

$$\frac{d}{dt}N(t) = (\epsilon - bN(t))\,N(t) \quad (N(t_0) = N_{ini}), \tag{2.41}$$

where ϵ, b are positive constants and $N_\infty := \frac{\epsilon}{b} > N_{ini}$ is the final number of ECs in the network. Equation (2.38) is regarded as an approximation of (2.41) when t is sufficiently small and $a = \epsilon N_{ini}$.

2.3.6 *Exact Solutions of the Continuous Model in Case of Constant VEGF Concentration*

Now we assume $\lambda_k^{(i)}(t) = 1$ for all (k, i), and the form of the function $F(n)$ as

$$F(n) = \begin{cases} v_0(n - n_e) & (n_e \le n < n_b) \\ 0 & (0 \le n < n_e) \end{cases}, \tag{2.42}$$

where a positive constant v_0 denotes the velocity of elongation. The density of ECs at time t, $\rho(t)$, satisfies

$$\rho(t_0) = \rho_0 := \frac{N_{ini}}{s_1 L_{ini}}. \tag{2.43}$$

Hence the effective number of ECs at the tip is given as

$$n_1(t_0) = n_0 := s_1 l \rho_0 = \frac{N_{ini} l}{L_{ini}}. \tag{2.44}$$

Without loss of generality, we can assume $n_e \le n_0 < n_b$. Because no elongation occurs if $n_0 < n_e$ and $L_1(t) = L_{ini}$, and it will start at a certain time t_0^* satisfying $n_1(t_0^*) = n_e$. We have only to replace t_0 with this t_0^*. Let t_1 be the time of first bifurcation. Since $n_1(t) = \frac{lN(t)}{L_1(t)}$, we have

$$\frac{d}{dt}\left(\frac{lN(t)}{n_1(t)}\right) = v_0(n_1(t) - n_e) \quad (t_0 < t < t_1). \tag{2.45}$$

By assuming that $N(t)$ satisfies (2.38) and that $at_0 = N_{ini}$ for simplicity, we have

$$N(t) = at. \tag{2.46}$$

From (2.45), (2.46), we have

$$\frac{la}{n_1(t)} - \frac{lat}{n_1(t)^2}\frac{d}{dt}n_1(t) = v_0(n_1(t) - n_e),$$

and

$$t\frac{d}{dt}n_1(t) = n_1(t) - \frac{v_0}{la}(n_1(t) - n_e)n_1(t)^2.$$

By using ξ, η given by

$$\xi - \eta = n_e, \quad \xi\eta = \frac{al}{v_0}, \tag{2.47}$$

we find

$$t\frac{dn_1}{dt} = -\frac{1}{\xi\eta}n_1(n_1 - \xi)(n_1 + \eta).$$

Integration over t gives

$$\int \frac{dt}{t} = \int \left\{ \frac{1}{n_1} - \left(\frac{\eta}{\xi + \eta} \right) \frac{1}{n_1 - \xi} - \left(\frac{\xi}{\xi + \eta} \right) \frac{1}{n_1 + \eta} \right\} dn_1.$$

Thus we obtain the following relation:

$$\frac{t}{t_0} = \left(\frac{n_1(t)}{n_1(t_0)} \right) \left| \frac{n_1(t_0) - \xi}{n_1(t) - \xi} \right|^{\frac{\eta}{\xi+\eta}} \left(\frac{n_1(t_0) + \eta}{n_1(t) + \eta} \right)^{\frac{\xi}{\xi+\eta}}. \tag{2.48}$$

Note that $\xi = \left(n_e + \sqrt{n_e^2 + 4al/v_0} \right)/2 > n_e$. From the above discussions, we conclude that if and only if it holds that

$$\xi > n_b, \tag{2.49}$$

bifurcation takes place and t_1 is given as

$$\frac{t_1}{t_0} = \left(\frac{n_b}{n_0} \right) \left| \frac{n_0 - \xi}{n_b - \xi} \right|^{\frac{\eta}{\xi+\eta}} \left(\frac{n_0 + \eta}{n_b + \eta} \right)^{\frac{\xi}{\xi+\eta}}, \tag{2.50}$$

and

$$L_1(t_1) = \frac{at_1 l}{n_b}. \tag{2.51}$$

In fact, if $\xi \le n_b$, n_1 decreases for $n_1(t_0) > \xi$ and increases for $n_1(t_0) < \xi$. Hence, in any case, it approaches to ξ and no bifurcation takes place. If $\xi > n_b$, t_1 is given by (2.50) because of (2.48). Since $N(t_1) = at_1$, we conclude that $L_1(t_1)$ is expressed as (2.51).

In order to investigate the behavior of elongation of blood vessels after bifurcation, it is useful to estimate the density of ECs $\rho(t)$:

$$\rho(t) = \frac{at}{\sum_k \sum_i s_k^{(i)} L_k^{(i)}(t)}. \tag{2.52}$$

Assume that no bifurcation takes place during $t_i < t < t_f$. From (2.29) and (2.52),

$$\frac{d}{dt} \frac{t}{\rho(t)} = V_m(\rho(t) - \rho_e). \tag{2.53}$$

Equation (2.53) can be solved for $\rho(t)$ in the same manners as those in (2.45). Then, we find

$$\frac{t_f}{t_i} = \left(\frac{\rho(t_f)}{\rho(t_i)} \right) \left(\frac{\rho(t_i) + \alpha}{\rho(t_f) + \alpha} \right)^{\frac{\alpha}{\alpha+\beta}} \left| \frac{\rho(t_i) - \beta}{\rho(t_f) - \beta} \right|^{\frac{\beta}{\alpha+\beta}}. \tag{2.54}$$

The positive constants $\alpha,\ \beta$ are defined by

$$\beta - \alpha = \rho_e, \quad \alpha\beta = \frac{1}{V_m}, \tag{2.55}$$

$$V_m := \frac{v_0 l \sum'(s_k^{(i)})^2}{a}, \quad \rho_e := \frac{n_e}{l} \frac{\sum' s_k^{(i)}}{\sum'(s_k^{(i)})^2} \tag{2.56}$$

where $\sum'$ denotes the summation over the pairs (k, i) that satisfy $n_k^{(i)} \geq n_e$.

As before, the (k, i) branch means the ith branch sprouting at a kth order bifurcation. If bifurcation takes place at the tip of (k^*, i^*) branch at time t^*, we find that

$$s_{k^*}^{(i^*)} = \max_k \left[\max_i \left[s_k^{(i)} \right] \right], \quad \rho(t^*) s_{k^*}^{(i^*)} l = n_b.$$

Here $\max_k \max_i$ is the maximum value of all the (k, i) branches that are extending.

The length of the branches can be calculated from (2.30), (2.54), and the following relation for arbitrary pair of branches (k, i), (p, j):

$$\frac{d}{dt}\left[s_p^{(j)} L_k^{(i)}(t) - s_k^{(j)} L_p^{(j)}(t) \right] = v_0 n_e (s_p^{(j)} - s_k^{(i)}), \tag{2.57}$$

which is deduced from (2.29). For example, in the 2nd-order bifurcation, let us suppose that $s_2^{(1)} > s_2^{(2)}$, and both (2,1) and (2,2) branches are extending for $t_1 < t < t_2^{(1)}$. At $t = t_2^{(1)}$, the 2nd-order bifurcation takes place and $\rho(t) s_2^{(1)} l = n_b$. Since $\rho(t_1) s_1 l = n_b$, (2.54) yields

$$t_2^{(1)} = \left(\frac{t_1 s_1}{s_2^{(1)}} \right) \left(\frac{n_b + \alpha s_1 l}{n_b + \alpha s_2^{(1)} l} \right)^{\frac{\alpha}{\alpha+\beta}} \left| \frac{n_b - \beta s_1 l}{n_b - \beta s_2^{(1)} l} \right|^{\frac{\beta}{\alpha+\beta}}.$$

Here $\alpha,\ \beta$ are given by

$$\beta - \alpha = \frac{n_e (s_2^{(1)} + s_2^{(1)})}{l \left\{ (s_2^{(1)})^2 + (s_2^{(1)})^2 \right\}}, \quad \alpha\beta = \frac{q}{v_0 l \left\{ (s_2^{(1)})^2 + (s_2^{(1)})^2 \right\}}.$$

Putting $L_1 = L_1(t_1)$, $L_2 = L_2^{(1)}(t_2^{(1)})$, $L_2' = L_2^{(2)}(t_2^{(1)})$, and noticing

$$\rho(t_2^{(1)}) = \frac{a t_2^{(1)}}{L_1 s_1 + s_2^{(1)} L_2 + s_2^{(2)} L_2'},$$

we find

$$L_1 s_1 + s_2^{(1)} L_2 + s_2^{(2)} L_2' = \frac{a t_2^{(1)} s_2^{(1)} l}{n_b}.$$

On the contrary, from (2.57),

$$s_2^{(2)} L_2 - s_2^{(1)} L_2' = (t_2^{(1)} - t_1) v_0 n_e (s_2^{(1)} - s_2^{(2)}).$$

Thus we have obtained simultaneous equations for unknown variables L_1, L_2. By solving these equations, we have

$$L_2 = \frac{1}{(s_2^{(1)})^2 + (s_2^{(1)})^2} \left\{ \left(\frac{a t_2^{(1)} s_2^{(1)} l}{n_b} - L_1 \right) s_2^{(1)} + (t_2^{(1)} - t_1) v_0 n_e (s_2^{(1)} - s_2^{(2)}) s_2^{(2)} \right\}.$$

The value of L_2' can be calculated in the same way.

General bifurcation time $t_k^{(i)}$ and branch length $L_k^{(i)}(t)$ can be obtained successively by similar estimations. When $s_{k+1}^{(i)} = s_{k+1}$ for any (k, i) branch, that is, all the cross sections of the branches arose at the same order bifurcation are the same, the kth order bifurcations take place at the same time t_k. The number of effective ECs $n_{k+1}^{(i)}(t)$ does not depend on i, and we put $n_{k+1}^{(i)}(t) = n_{k+1}(t)$. Then, we obtain the following relations:

$$\frac{t_k}{t_{k-1}} = \left(\frac{s_{k-1}}{s_k} \right) \left(\frac{n_b + \alpha_k l s_k}{n_b + \alpha_k l s_{k-1}} \right)^{\frac{\alpha_k}{\alpha_k + \beta_k}} \left| \frac{\beta_k l s_k - n_b}{\beta_k l s_{k-1} - n_b} \right|^{\frac{\beta_k}{\alpha_k + \beta_k}}, \tag{2.58}$$

where $s_0 := \frac{n_b}{\rho(t_0) l}$, and we assume that $n_k(t_{k-1}) = \rho(t_{k-1}) s_k l \geq n_e$ for simplicity. If $n_k(t_{k-1}) < n_e$, denoting by V_k the total volume of the blood vessel network at t_k, we have only to replace t_{k-1} with $t_{k-1}^* := \frac{n_e V_k}{a s_k l}$ in (2.58). The constants α_k, β_k are given by

$$\beta_k - \alpha_k = \frac{n_e}{l s_k}, \quad \alpha_k \beta_k = \frac{a l}{2^{k-1} v_0 s_k^2}. \tag{2.59}$$

If we denote by L_{k+1} the length of the branches generated at the kth order bifurcations, we find

$$L_1 = \frac{a l t_1}{n_b}, \qquad L_{k+1} = \frac{a l (t_{k+1} - t_k)}{2^k n_b} \quad (k = 1, 2, \ldots). \tag{2.60}$$

In case $N(t) = N_{ini} + a(t - t_0)$, the number of bifurcations are finite. We note that $s_{k+1} < s_k$ $(k = 1, 2, \ldots)$. For $t \geq t_{k-1}$, (2.53) yields

$$t\frac{d}{dt}\rho(t) = -\rho(t)(\rho(t) - \beta_k)(\rho(t) + \alpha_k).$$

Hence $\rho(t)$ monotonically approaches to the value β_k, and no bifurcation takes place if $ls_k\beta_k \leq n_b$. Since

$$\beta_k = \frac{n_e + \sqrt{n_e^2 + (4al/2^{k-1}v_0)}}{2ls_k},$$

a branch bifurcates when the inequality

$$n_e + \sqrt{n_e^2 + (4al/2^{k-1}v_0)} > 2n_b$$

is satisfied. Thus if the inequality

$$\frac{al}{v_0 n_b(n_b - n_e)} > 2^{k-1}$$

is satisfied, the kth order bifurcation can occur. Therefore, if the kth order bifurcation is the last bifurcation, an integer k satisfies the inequality

$$k - 1 < \log_2\left(\frac{al}{v_0 n_b(n_b - n_e)}\right) \leq k.$$

This value is given as

$$\left\lceil \log_2\left(\frac{al}{v_0 n_b(n_b - n_e)}\right)\right\rceil, \tag{2.61}$$

where $\lceil x \rceil$ is the smallest integer that is equal to or greater than x.

When we take the cell division of ECs in the blood vessel network, we have

$$N(t) = e^{\epsilon t} N_0 \qquad (N_0 > 0). \tag{2.62}$$

In this case, a blood vessel continues to bifurcate up to infinity. Practically, the supply of ECs will stop within a finite time interval, and the elongation and bifurcation of blood vessels also finish. The function $N(t)$ could be expressed by (2.41) or some nonlinear evolution equation. In these cases, analysis for angiogenesis becomes a little complicated but can be done fairly easily with numerical estimations.

2.3.7 Numerical Simulations in the Presence of VEGF Concentration Gradient

In most experimental situations, the system for the observation of angiogenesis is essentially two spatial dimensions. We performed numerical simulations of our continuous model under the following conditions:

- Equation (2.29) gives the time evolution of blood vessels.
- Supply and cell division of ECs are given by (2.41).
- Elongation and bifurcation of blood vessels take place in two spatial dimensions.
- The cross sections of new branches are determined according to the Murray's law ($m = 1.5$).
- The effects of concentration gradient of VEGF are incorporated to $\lambda_k^{(i)}(t)$, which is as follows.

  ▷ The concentration of VEGF is constant in time and is denoted by $f(x, y)$ at the position (x, y). The concentration gradient is gentle, and $f(x, y)$ is almost constant in a branch.

  ▷ The first blood vessel extends from the origin toward the direction of $x > 0$ along x-axis.

  ▷ We determine a unit length Δt. For the initial blood vessel,

  $$\Lambda_1 := \int_0^{\Delta t} f(x, 0)\, dx,$$

  and we put $\lambda_1(t) = c\Lambda_1$, where c is a positive constant.

  ▷ The 1st bifurcation takes place at time $t = t_1$, and the position $(x, y) = (L(t_1), 0)$. We assume that the angle between the new branches is 60°, and those between the original vessel are both 150°.

  ▷ By integrating $f(x, y)$ along the new branch direction by the unit length Δx, we determine the activity $\lambda_2^{(1)}(t)$, $\lambda_2^{(2)}(t)$ as earlier. Namely, we perform the line integrals of $f(x, y)$ from $(L(t_1), 0)$ to $(L(t_1) + \frac{\sqrt{3}}{2}\Delta x, \frac{1}{2}\Delta x)$ and to $(L(t_1) + \frac{\sqrt{3}}{2}\Delta x, -\frac{1}{2}\Delta x)$ and denote the values by $\Lambda_2^{(1)}$ and $\Lambda_2^{(2)}$, respectively. Then, $\lambda_2^{(1)}(t) = c\Lambda_2^{(1)}$, $\lambda_2^{(2)}(t) = c\Lambda_2^{(2)}$.

  ▷ At the higher order bifurcations, we perform line integrals along the direction of new branches and determine $\lambda_k^{(i)}(t)$ in the same manner.

Typical numerical results are shown in Figs. 2.31 and 2.32. Both figures show the final blood vessel networks after the supply of ECs finishes. We find that the density of vessels is high, while average length of branches is small in the area where VEGF concentration is high.

Figure 2.33 shows the dependence of the threshold value n_e. As n_e becomes larger, the times of bifurcation increase and length of a branch becomes shorter.

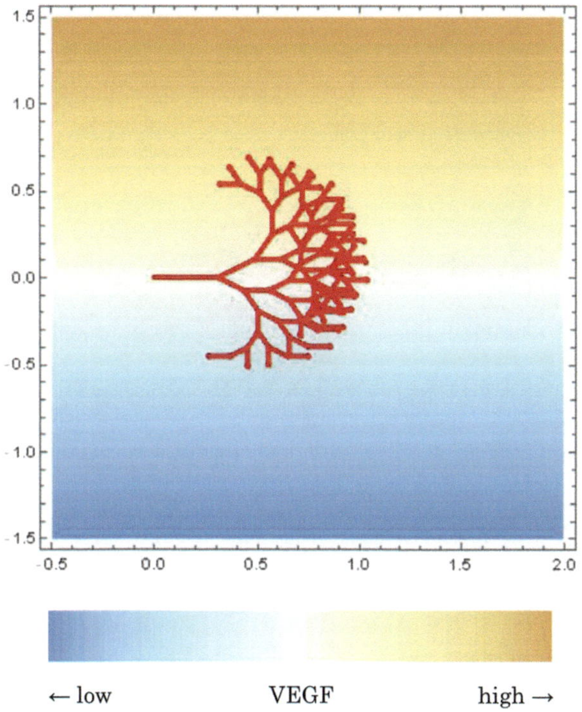

Fig. 2.31 Bifurcation of neogenetic blood vessels under the effect of VEGF. The function $f(x, y) = 1 + (\tan^{-1} y)/\pi$. The other parameters are the same as those in Fig. 2.30

Figure 2.34 shows the change of patterns according to the change of VEGF concentration. Higher concentration results in the increase of bifurcation and shortening of branch length.

Figure 2.35 shows the change of number of branches as the change of VEGF concentration and that of the threshold value n_e. Finally, we show the dependence of branch length on the VEGF concentration in Fig. 2.36. The length of branches becomes shorter as VEGF concentration becomes higher. While it reduces for the first several bifurcations but turns to increase later.

2.3.8 *Concluding Remarks*

In conclusion, we have presented a discrete dynamical model for angiogenesis that successfully reproduces the cell mixing behavior, elongation, and bifurcation. The dynamics of ECs is supposed to be mainly ruled by deterministic two-body interactions that consist of short-range repulsion due to excluded volume effect and long-range attractive force through pseudopodia. If ECs are supplied to the origin at

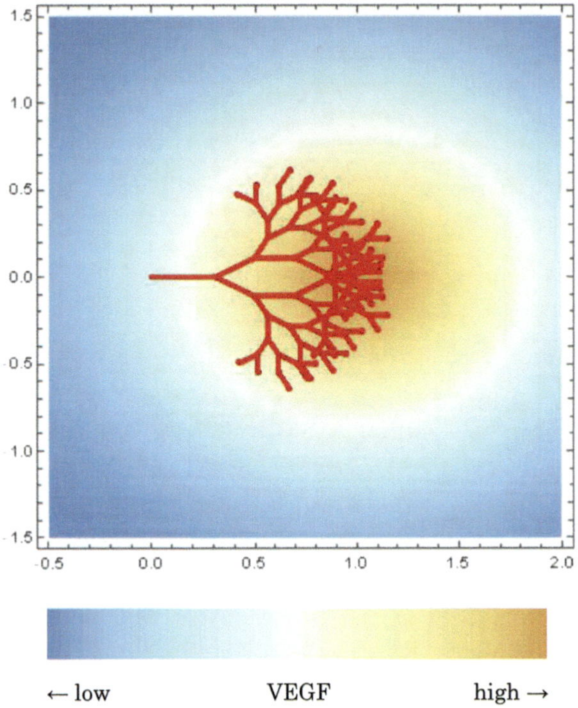

Fig. 2.32 Bifurcation of neogenetic blood vessels under the effect of VEGF. The function $f(x, y) = 0.8 + 0.5 \exp[-(x-1)^2 - y^2]$. The other parameters are the same as those in Fig. 2.30

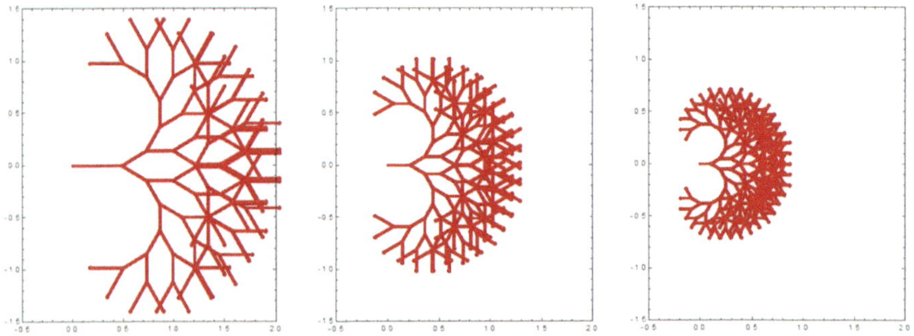

Fig. 2.33 The parameter n_e dependence of bifurcation patterns of neogenetic vessels. The values are $n_e = 0.1n_b$, $0.5n_b$, and $0.9n_b$, respectively, from left. The other parameters are the same as those in Fig. 2.30

a constant rate (a = const.), then the system is completely deterministic. However, if we assume that ECs are supplied with a given probability (equivalently a given mean value $\langle a \rangle$), then the injection interval becomes a stochastic variable, although the qualitative behavior of the model is not affected by the way of injection. Under

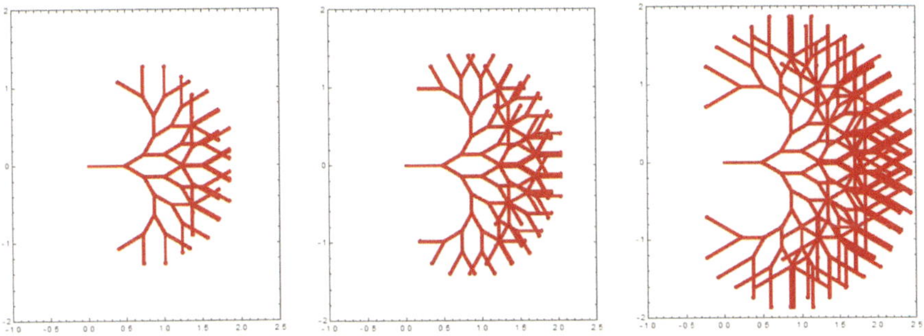

Fig. 2.34 VEGF dependence of bifurcation patterns of neogenetic vessels. The concentration of VEGF is spatially uniform and the values of $f(x, y) = 0.4$, 1.0, and 1.4, respectively, from left, and $n_e = 0.1n_b$. The other parameters are the same as those in Fig. 2.30

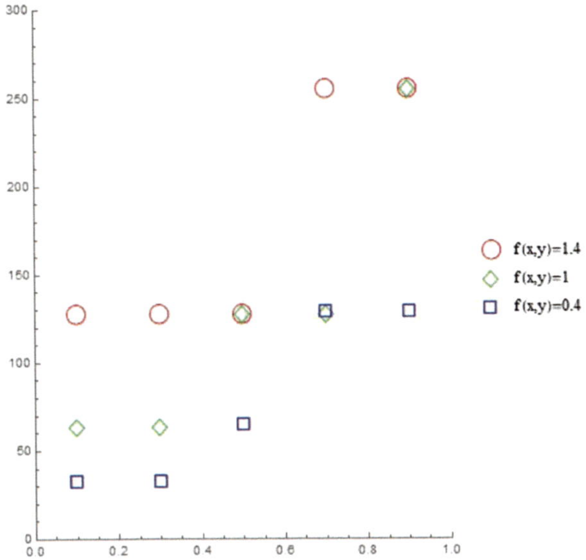

Fig. 2.35 The number of bifurcations with respect to the n_e/n_b and $f(x, y)$. (Here $f(x, y)$ is a constant function.) The other parameters are the same as those in Fig. 2.30

this interaction, the tip position of neogenetic vessel at time step t, $\ell(t)$, develops as $\ell(t) \propto t^{2/3}$. We have given an interpretation on this value of exponent, 2/3, based on the equation of continuity and a hypothesis of the existence of a scaling function. The oscillating bound states in case of mere repulsive interaction were also presented, which would give a reasoning for the cell-mixing effects in purely repulsive interactions. The behavior of ECs and evolution patterns of new vessels in early stage are well-reproduced by the present model. Our main conclusion in the present research is that deterministic two-body interaction between ECs

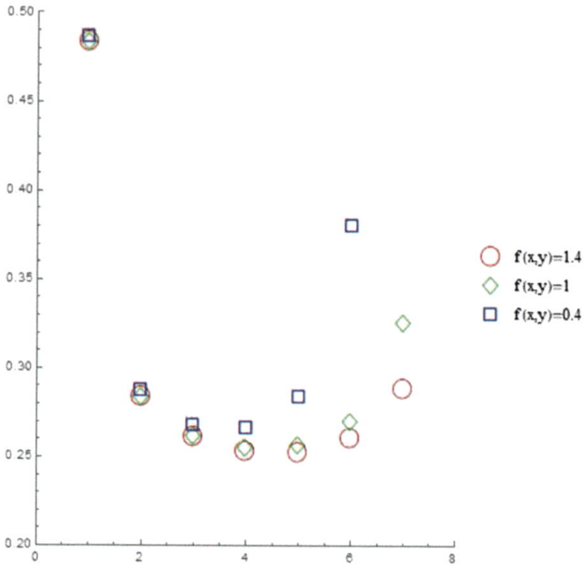

Fig. 2.36 VEGF dependence of the lengths of the branches L_k generated by $k - 1$th bifurcation ($k = 1, 2, \ldots, 7$). The parameters are the same as those in Fig. 2.34

can give rise to cell-mixing effect, elongation and bifurcation, which suggests the importance of further investigation for interactive behavior of ECs in angiogenesis. However, since we do not incorporate recombination of vessels and cell division, the time evolution patterns for long time span deviate from actual networks of blood vessels. In particular, the length of vessels tends to be longer with increase of bifurcations, which is not observed in actual systems. We conjecture that this discrepancy is caused by the effect of cell division that was rarely observed in the time-lapse imaging experiments but will be important for long time span. In the models, we have not, however, included chemotaxis, a gradient distribution of VEGF and remodeling of blood vessels that are important in construction of in vivo blood vessel networks. Incorporating these factors and closely examining forthcoming experimental results, we wish to develop the present models so that we can quantitatively explain various types of angiogenetic phenomena and provide a theoretical framework for clinical trials targeting angiogenesis.

2.4 Two-Dimensional Pattern Formation with Ellipses

As we have seen in the previous Sect. 2.3, the neogenetic vessels are configured through collective cell migration such as cell mixing. Our one-dimensional models also show the necessity of the attractive and repulsive forces for neogenetic development of vessels. In [60], the range of distance-dependent intercellular forces are estimated under the assumption of two-body interactions; a repulsive force in $\sim\ 8\mu m$ and attractive force in $8\ \sim\ 20\mu m$ in the sprouting branch from in vitro mouse aortic sheet. The long-range attractive force mainly originates from oncoming neighbor cells, which results from adhesion (cf. Sect. 2.1.4). On the contrary, the short-range repulsive force is considered to be the results of excluded volume effect. In this section, we consider a two-dimensional discrete model with excluded volume effect, in which the cells are expressed not as points but as ellipses that have finite volume. It has been suggested that angiogenic sprouting is a natural emergent property of elongated, adhesive objects in a stochastic system[43]. We investigate a discrete model in deterministic manner without optimization such as simulated annealing.

2.4.1 *Two-Dimensional Extension of the Model*

Suppose each EC is approximated as ellipse. Let us define the shape of ith EC at time step t by the length of semi-major (resp., semi-minor) axis a_i^t (resp., b_i^t) and the rotation angle of major axis $\varphi_i^t \in [0, \pi)$ (Fig. 2.37). For the center position ${}^t(x_i^t, y_i^t)$ of ith EC, we consider a two-dimensional extension of the (2.17) and (2.18) in Sect. 2.3.2:

$$\begin{aligned} \mathbf{z}_i^{t+1} - \mathbf{z}_i^t &= \mathbf{v}_i^t, \\ \mathbf{v}_i^{t+1} - \mathbf{v}_i^t &= -\gamma \mathbf{v}_i^t + \sum_{j \neq i} \mathbf{F}_{i,j}^t, \end{aligned} \tag{2.63}$$

where $\mathbf{z}_i^t = {}^t(x_i^t, y_i^t) \in \mathbb{R}^2$ is the position of the ith EC at time step $t \in \mathbb{Z}_{\geq 0}$ and $\mathbf{v}_i^t =: {}^t(v_{x,i}^t, v_{y,i}^t) \in \mathbb{R}^2$ the velocity, and left upper t stands for transpose.

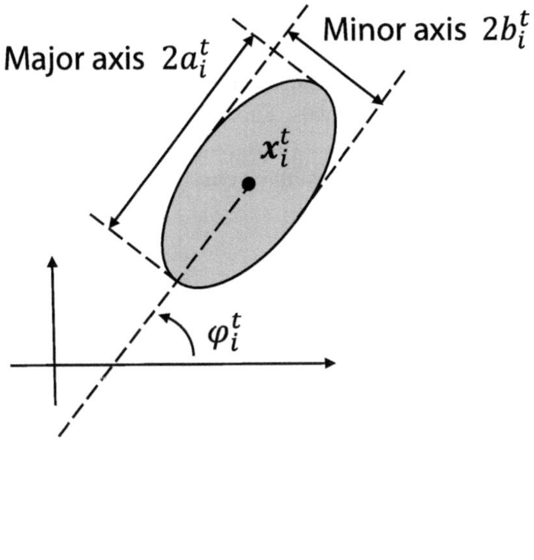

Fig. 2.37 Sketch of ith EC at time step t, whose coordinate is x_i^t, semi-major axis a_i^t, semi-minor axis b_i^t, and rotation angle φ_i^t

Let us define the interaction between two ellipses. Parametric representation of the solid ellipse of ith EC by $0 \leq r \leq 1$ with

$$\begin{pmatrix} x \\ y \end{pmatrix} = \mathbf{z}_i^t + R(\varphi_i^t) \begin{pmatrix} a_i^t & 0 \\ 0 & b_i^t \end{pmatrix} \begin{pmatrix} r\cos\theta \\ r\sin\theta \end{pmatrix} =: g_i(r, \theta),$$

where R is the rotating matrix, $r \in [0, 1]$ the radius, and $\theta \in [0, 2\pi)$ the parameter of the curve. Let E_i be the domain of the ith ellipse,

$$E_i := \{g_i(r, \theta)\}_{0 \leq r \leq 1, 0 \leq \theta < 2\pi} . \tag{2.64}$$

On the boundary ($r = 1$) of E_i, we select the D sampling points by $g_i(1, \theta_k)$, where $\theta_k = 2\pi k/D$, $k \in \{0, 1, \ldots, D-1\}$, and $D \in \mathbb{Z}_{>0}$ is the common parameter for all ECs. Let B_i the set of the D sampling points on the boundary of E_i, that is,

$$B_i := \{g_i(1, \theta_k)\}_{k \in \{0,1,\ldots,D-1\}} . \tag{2.65}$$

The two-body interaction causes the force and rotation depending on positions and directions of two ECs as follows (Figs. 2.38, 2.39, and 2.40):

Repulsive force: ECs are not solid objects but may not share the same space, and the nucleus inside EC is also known to be rigid[52]. If two ECs are close enough, they repel each other. For each point $\mathbf{q}$ that satisfies $\mathbf{q} \in B_i$ and $\mathbf{q} \in E_j$, accumulate the repulsive force $f_r \mathbf{e}_{i,j}^t$ to ith EC. In other words, if a sampling point $\mathbf{q} \in B_i$ of the ith EC lies inside jth EC, ith EC receives the repulsive force along the centers.

Attractive force: Since ECs are apart from each other without attractive forces, wider range of attractive force than repulsive one may be indispensable. We assume isotropic attractive force $-f_a \mathbf{e}_{i,j}^t$ for ith EC along to jth EC if the

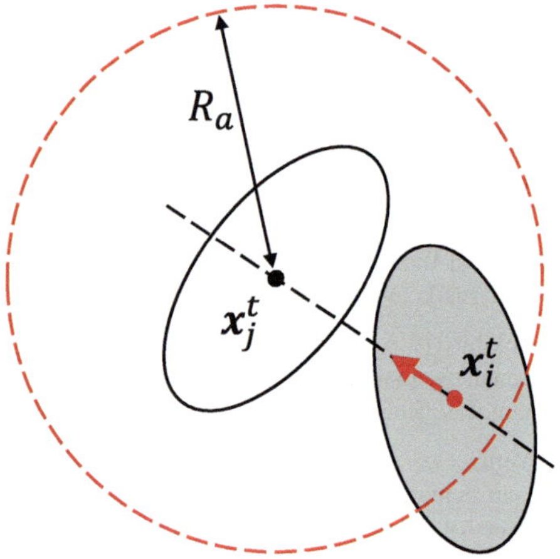

Fig. 2.38 Sketch of attractive force to ith EC (red arrow). If the center of ith EC locates in the disk centered at jth EC with radius R_a, the ith EC (gray ellipse) receives attractive force along to jth EC

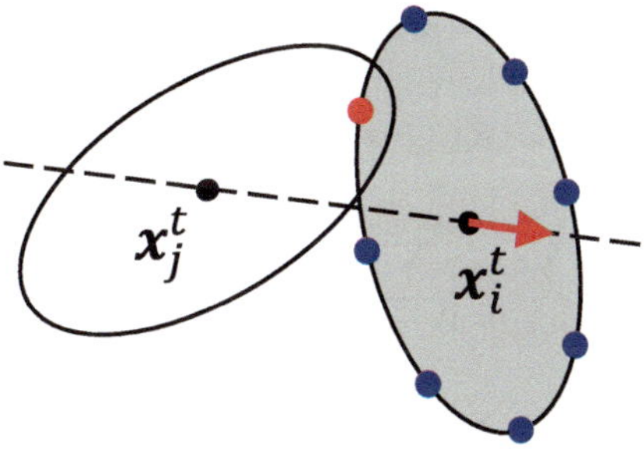

Fig. 2.39 Sketch of repulsive force to ith EC (red arrow). We show the case of $D = 8$ as an example. Eight sampling points (blue and red points) are depicted on the boundary of ith EC. In this case, overlap is detected since a sampling point of ith EC (red circle) is inside the jth EC. The ith EC (gray ellipse) suffers repulsive force away from jth EC (red arrow)

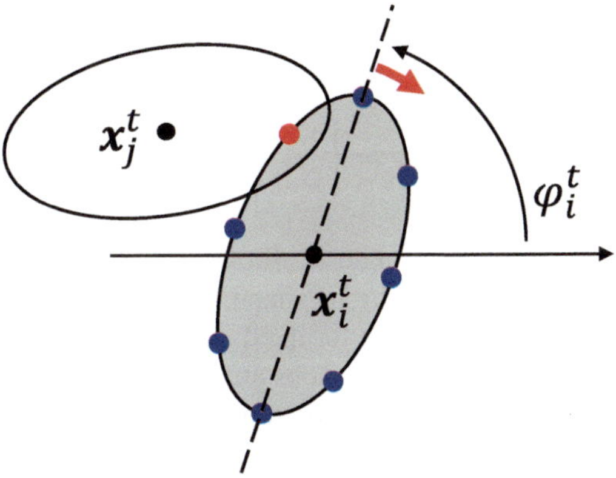

Fig. 2.40 Sketch of rotating force to ith EC (red arrow). A sampling point of ith EC (red circle) located in the jth EC, so the ith EC (gray ellipse) rotates to avoid overlap. We show the case of $D = 8$ as an example, in which the red point corresponds to $k = 1$ and $\theta_k = \pi/4$ in the text. Therefore, φ_i^t of the ith EC decrease f_p by this overlap; that is, ith EC rotates clockwise

distance between two centers $\|\mathbf{z}_i^t - \mathbf{z}_j^t\|$ is less than R_a. Since this attractive force and the above repulsive one act on the same line, we also assume $f_a < f_r$ so as not to counteract the repulsive force.

Rotating force: Similar to the above repulsive force, the excluded volume effect prevents the overlap of ECs. If the sampling point $\mathbf{q}$ of the EC i exists inside jth EC, ith EC rotates for avoiding overlap.

Thus, we obtain the two-body interaction as follows:

$$\begin{aligned}\mathbf{F}_{i,j}^t &= -f_a \mathbf{e}_{i,j}^t \Theta(R_a - \|\mathbf{z}_i^t - \mathbf{z}_j^t\|) + \sum_{\mathbf{q}\in B_i} f_r \mathbf{e}_{i,j}^t \mathbf{1}_{E_j}(\mathbf{q}), \\ \varphi_i^{t+1} - \varphi_i^t &= - \sum_{k\in\{0,1,\ldots,D-1\}} \sum_{j\neq i} f_p \sin(2\theta_k) \mathbf{1}_{E_j}(g_i(1,\theta_k)),\end{aligned} \tag{2.66}$$

where $\mathbf{e}_{i,j}^t := (\mathbf{z}_i^t - \mathbf{z}_j^t)/\|\mathbf{z}_i^t - \mathbf{z}_j^t\|$, that is, the unit vector along the relative center positions of two ECs, and f_a, f_r and f_p are the common parameters for all ECs, and Θ the step function

$$\Theta(x) := \begin{cases} 1 & (x > 0) \\ 0 & (x \leq 0) \end{cases},$$

and $\mathbf{1}_A$ the characteristic function on a set A

$$\mathbf{1}_A(x) := \begin{cases} 1 & (x \in A) \\ 0 & (x \notin A) \end{cases}.$$

Note that we defined the rotation angle of ellipse as $0 \leq \varphi_i^t < \pi$; therefore, we do not distinguish the polarity of EC by its shape, that is, the direction of migration (Fig. 2.3). The velocity $\mathbf{v}_i^t$ plays a role for the direction of migration.

For simplicity, we limit ECs to the constant and common shapes $a_i^t = a$ and $b_i^t = b$ for all t and i ($a \geq b$), and furthermore choose the area of ellipse as unit $ab = 1$ for length scaling. Thus, the shapes of all ECs are characterized by only one parameter; the flattening $f := 1 - b/a = 1 - 1/a^2$.

2.4.2 Pattern Formation and Order Parameter

Let us suppose that the case ECs are regularly supplied to the system. Assume that the ECs are periodically injected with the interval 10 and uniformly at random with the angle $[0, \pi)$ in the unit circle (radius one) at the fixed point. In Fig. 2.41, we show the numerical results of (2.63) with various flattenings. The figures are snapshots at $t = 5000$; therefore, 500(=5000/10) ECs exist on each plane as a result. For $f = 0$ case (leftmost), the ECs form a rounded compact mass. In this $f = 0$ case, all forces are isotropic and independent of the direction φ_i^t because all ECs are circle. On the contrary for larger flattening, the patterns appear to be sprouting to peripheral area. In Fig. 2.42, we show the number of square boxes required to cover the set of the center positions of all ECs on the plane with respect to the grid spacing. By fitting the curve, the box-counting dimensions are estimated as $\sim$ 1.7, 1.6, 1.2 for $f = 0, 0.3, 0.7$, respectively; the more ECs elongate, the more branches grow one dimensionally. Thus, this change originates in only flattening of ellipses. Since the

Fig. 2.41 Patterns formed by ECs on the plane at $t = 5000$. The flattening f is 0, 0.3, and 0.7, respectively, from left to right. Common parameters are $D = 16$, $R_a = 5$, $f_a = 0.005$, $f_r = 0.015$, $f_p = 0.005$, and $\gamma = 0.1$. ECs are regularly injected at the center of each figure

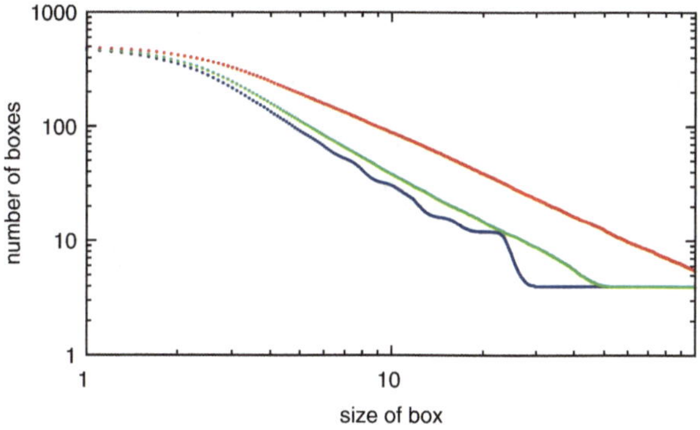

Fig. 2.42 Box-counting dimensions are estimated: dependencies of the number of occupied boxes with respect to the size of boxes at the snapshot $t = 5000$. Each line is averaged by 100 trials (Blue: $f = 0$, Green: $f = 0.3$, Red: $f = 0.7$). Other parameters are the same as Fig. 2.41. Estimated box-counting dimensions are 1.7, 1.6, 1.2, respectively

elongated ellipse frequently collides at the front part compared to side part, the major axes are aligned with each other. Note that our model is deterministic, and the randomness has to do with the way of the injection at the fixed center. Though the pattern radiating in all directions results from this injection, the slender branches in Fig. 2.41 are intrinsic to our model. In Figs. 2.43, 2.44, and 2.45, we show the time evolutions from initial state consisting of scattered ECs in a square region. The three initial states (leftmost) are prepared from the pseudorandom number generator with the same seed. Fig. 2.43 shows that the circles ($f = 0$) form a few rounded compact mass due to isotropic attractive and repulsive forces. The larger the flattening f becomes, the more the branches elongate (Figs. 2.44 and 2.45). The center figure in Fig. 2.45 shows that the mesh structure, which is similar to retina pattern, is temporarily formed; however, such structure disappears as time goes on. In our model, we do not take extracellular matrix into account; that is, ECs are not binded

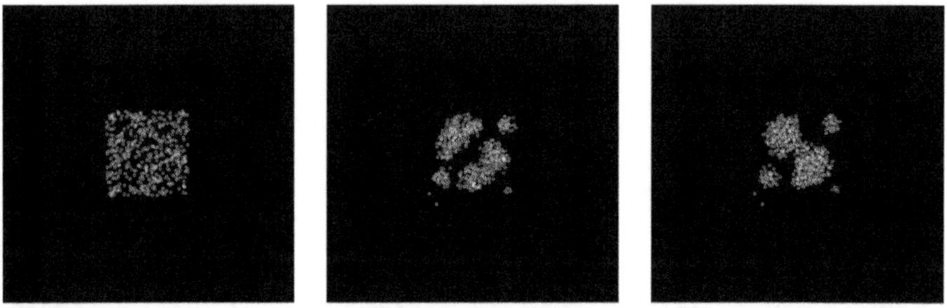

Fig. 2.43 Time evolution of 500 ECs on the plane for the flattening $f = 0$ without injection. Each figure is the snapshot at $t = 1, 1500$, and 5000, respectively, from left to right. Parameters are the same as Fig. 2.41

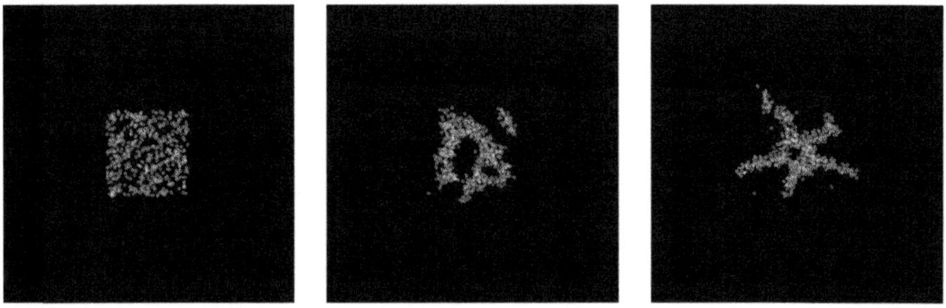

Fig. 2.44 Time evolution of 500 ECs on the plane for the flattening $f = 0.3$ without injection. Each figure is the snapshot at $t = 1, 1500$, and 5000, respectively, from left to right. Parameters are the same as Fig. 2.41

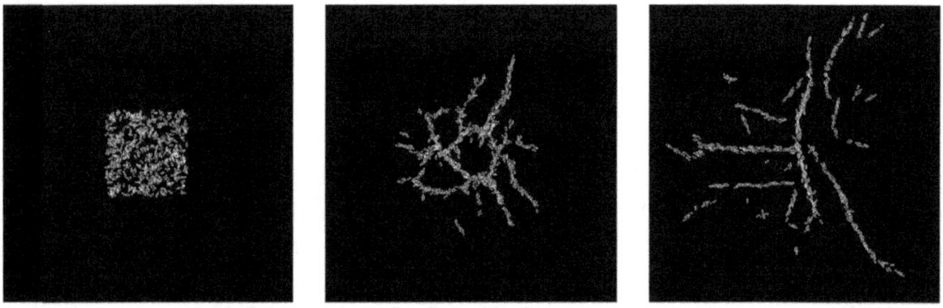

Fig. 2.45 Time evolution of 500 ECs on the plane for the flattening $f = 0.7$ without injection. Each figure is the snapshot at $t = 1, 1500$, and 5000, respectively, from left to right. Parameters are the same as Fig. 2.41

to matrix through focal adhesion. Stabilizing these structures in our model will be a future problem.

In one branch, ECs tend to be aligned along each other. To quantify the alignment, let us define the local order parameter as follows. Suppose two ECs with φ_i and φ_j, then the parallel (resp., perpendicular) ECs give $\cos 2(\varphi_i - \varphi_j) =$

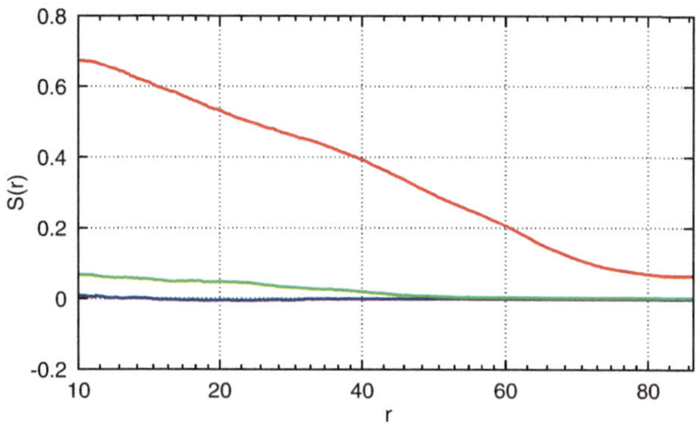

Fig. 2.46 Order parameters $S^t(r)$ at $t = 5000$ for the patterns in Figs. 2.43, 2.44, and 2.45. Blue: $f = 0$, Green: $f = 0.3$, and Red: $f = 0.7$

1(resp. -1). Averaging $\cos 2(\varphi_i - \varphi_j)$ over jth EC around ith EC, we may obtain a locally defined order for each EC:

$$S_i^t(r) := \frac{1}{N_i^t(r)} \sum_{j \neq i, \|\mathbf{z}_i^t - \mathbf{z}_j^t\| \leq r} \cos 2(\varphi_i^t - \varphi_j^t),$$

where $N^t(r)$ is the number of ECs around ith EC at time t. In addition, averaging $S_i^t(r)$ over all ECs yields the local order parameter of the system $S^t(r) := \langle S_i^t(r) \rangle_i$. Note that this order parameter $S^t(r)$ may be negative, though $S^t(r)$ gives zero for randomly oriented ECs. In Fig. 2.46, we show the distance dependency on the order parameter for the patterns in Figs. 2.43, 2.44, and 2.45 at $t = 5000$. Since the circle shape ($f = 0$) causes random rotating regardless of the ECs direction, the order parameter is always almost zero, whereas for nonzero f, it may be seen that the order parameter of smaller r is large. Rough estimation from the semi-log figure gives the exponential decay; therefore, it is supposed that the direction of EC has the typical correlation scale, which depends on the parameters f_a, f_r, f_p, and the flattening f. This correlation length is considered to be a possible factor that determines the above mesh structure. Analysis of the bifurcation of neogenetic blood vessels from the "dynamical systems" point of view could shed new light on these issues.

References

1. Abraham, S., Yeo, M., Montero-balaguer, M., Paterson, H., Dejana, E., Marshall, C.J.: Report interaction suppresses sprouting via signaling to MCL2 phosphorylation. Curr. Biol. **19**(8), 668–674 (2009)
2. Anderson, A.R.A., Chaplain, M.J.: Continuous and discrete mathematical models of tumor-induced angiogenesis. Bull. Math. Biol. **60**, 857–900 (1998)

3. Anderson, A.R.A., Chaplain, M.A.J., Rejiniak, K.A. (eds.): Single-Cell-Based Models in Biology and Medicine. Mathematics and Biosciences in Interaction Series. Birkhäuser, Basel (2007)
4. Arima, S., Nishiyama, K., Ko, T., Arima, Y., Hakozaki, Y., Sugihara, K., Koseki, H., Uchijima, Y., Kurihara, Y., Kurihara, H.: Angiogenic morphogenesis driven by dynamic and heterogeneous collective endothelial cell movement. Development **138**, 4763–4776 (2011)
5. Arima, S., Nishiyama, K., Ko, T., Arima, Y., Hakozaki, Y., Sugihara, K., et al.: Angiogenic morphogenesis driven by dynamic and heterogeneous collective endothelial cell movement. Development **138**(21), 4763–4776 (2011)
6. Bauer, A.L., Jackson, T.L., Jiang, Y.: A cell-based model exhibiting branching and anastomosis during tumor-induced angiogenesis. Biophys. J. **92**, 3105–3121 (2007)
7. Bauer, A.L., Jackson, T.L., Jiang, Y.: Topography of extracellular matrix mediates vascular morphogenesis and migration speeds in angiogenesis. PROS Compt. Biol. **5**, e1000445 (2009)
8. Bentley, K., Franco, C.A., Philippides, A., Blanco, R., Dierkes, M., Gebala, V., et al.: The role of differential VE-cadherin dynamics in cell rearrangement during angiogenesis. Nat. Cell Biol. **16**(4), 309–321 (2014)
9. Blanchoin, L., Boujemaa-paterski, R., Sykes, C., Plastino, J.: Actin dynamics, architecture, and mechanics in cell motility. Physiol. Rev. **94**(1), 253–263 (2014)
10. Brasch, J., Harrison, O.J., Ahlsen, G., Carnally, S.M., Henderson, R.M., Honig, B., et al.: Structure and binding mechanism of vascular endothelial cadherin: A divergent classical cadherin. J. Mol. Biol. **408**(1), 57–73 (2011)
11. Cao, J., Ehling, M., März, S., Seebach, J., Tarbashevich, K., Sixta, T., et al.: Polarized actin and VE-cadherin dynamics regulate junctional remodelling and cell migration during sprouting angiogenesis. Nat. Commun. **8**(1), 2210 (2017)
12. Carmeliet, P., Dor, Y., Herbert, J.M., Fukumura, D., Brusselmans, K., Dewerchin, M., et al.: Role of HIF-1α in hypoxia-mediated apoptosis, cell proliferation and tumour angiogenesis. Nature **395**(6692), 485–490 (1998)
13. Carmeliet, P., Lampugnani, M.G., Moons, L., Breviario, F., Compernolle, V., Oise Bono, F., et al.: Targeted deficiency or cytosolic truncation of the VE-cadherin gene in mice impairs VEGF-mediated endothelial survival and angiogenesis. Cell **98**, 147–157 (1999)
14. Chung, A.S., Lee, J., Ferrara, N.: Targeting the tumour vasculature: insights from physiological angiogenesis. Nat. Rev. Cancer **10**(7), 505—514 (2010)
15. Daub, J.T., Merks, M.H.: A cell-based model of extracellular-matrix-guided endothelial cell migration during angiogenesis. Bull. Math. Biol. **75**, 1377–1399 (2013)
16. Ferrara, N., Hillan, K.J., Gerber, H., Novotny, W.: Discovery and development of bevacizumab, an anti-VEGF antibody for treating cancer. Nat. Rev. Drug Discov. **3**(5), 391–400 (2004)
17. Friedl, P., Gilmour, D.: Collective cell migration in morphogenesis, regeneration and cancer. Nat. Rev. Mol. Cell Biol. **10**(7), 445–457 (2009)
18. Friedl, P., Weigelin, B.: Interstitial leukocyte migration and immune function. Nat. Immunol. **9**(9), 960–969 (2008)
19. Gaengel, K., Niaudet, C., Hagikura, K., Laviña, B., Muhl, L., Hofmann, J.J., et al.: The sphingosine-1-phosphate receptor S1PR1 restricts sprouting angiogenesis by regulating the interplay between VE-cadherin and VEGFR2. Develop. Cell **23**(3), 587–599 (2012)
20. Gamba, A., Ambrosi, D., Coniglio, A., de Candia, A., DiTalia, S., Giraudo, E., Serini, G., Preziosi, L., Bussolino, F.: Percolation, morphogenesis, and burgers dynamics in blood vessels formation. Phys. Rev. Lett. **90**(11), 118101 (2003)
21. Garrett, T.A., Buul, J.V., Burridge, K.: VEGF-induced rac1 activation in endothelial cells is regulated by the guanine nucleotide exchange factor VAV2. Exp. Cell Res. **313**(15), 3285–3297 (2007)
22. Gavard, J., Gutkind, J.S.: VEGF controls endothelial-cell permeability by promoting the β-arrestin-dependent endocytosis of VE-cadherin. Nat. Cell Biol. **8**(11), 1223–1234 (2006)
23. Gerhardt, H., Golding, M., Fruttiger, M., Ruhrberg, C., Lundkvist, A., Abramsson, A., et al.: VEGF guides angiogenic sprouting utilizing endothelial tip cell filopodia. J. Cell Biol. **161**(6), 1163–1177 (2003)

24. Gory-Fauré, S., Prandini, M.H., Pointu, H., Roullot, V., Pignot-Paintrand, I., Vernet, M., et al.: Role of vascular endothelial-cadherin in vascular morphogenesis. Development **126**(10), 2093–2102 (1999)
25. Hellström, M., Phng, L.K., Hofmann, J.J., Wallgard, E., Coultas, L., Lindblom, P., et al.: Dll4 signalling through Notch1 regulates formation of tip cells during angiogenesis. Nature **445**(7129), 776–780 (2007)
26. Hurwitz, H., Fehrenbacher, L., Novotny, W., Cartwright, T., Hainsworth, J., Heim, W., et al.: Bevacizumab plus irinotecan, fluorouracil, and leucovorin for metastatic colorectal cancer. N. Engl. J. Med. **350**(23), 2335–2342 (2004)
27. Jain, T., Nikolopoulou, E.A., Xu, Q., Qu, A.: Hypoxia inducible factor as a therapeutic target for atherosclerosis. Pharmacol. Therapeut. **183**, 22–33 (2018)
28. Jakobsson, L., Franco, C.A., Bentley, K., Collins, R.T., Ponsioen, B., Aspalter, I.M., et al.: Endothelial cells dynamically compete for the tip cell position during angiogenic sprouting. Nat. Cell Biol. **12**(10), 943–953 (2010)
29. Karaman, S., Leppänen, V.M., Alitalo, K.: Vascular endothelial growth factor signaling in development and disease. Development **145**(14), dev151019 (2018)
30. Kerbel, R.S.: Tumor angiogenesis. N. Engl. J. Med. **358**(19), 2039–2049 (2008)
31. Khan, K.A., Kerbel, R.S.: Improving immunotherapy outcomes with anti-angiogenic treatments and vice versa. Nat. Rev. Clin. Oncol. **15**(5), 310–324 (2018)
32. Komarova, Y.A., Kruse, K., Mehta, D., Malik, A.B.: Protein interactions at endothelial junctions and signaling mechanisms regulating endothelial permeability. Circ. Res. **120**(1), 179–206 (2017)
33. Lampugnani, M., Resnati, M., Raiteri, M., Pigott, R., Pisacane, A., Houen, G., et al.: A novel endothelial-specific membrane protein is a marker of cell-cell contacts. J. Cell Biol. **118**(6), 1511–1522 (1992)
34. Lampugnani, M.G., Zanetti, A., Corada, M., Takahashi, T., Balconi, G., Breviario, F., et al.: Contact inhibition of VEGF-induced proliferation and the phosphatase DEP-1/CD148. J. Cell Biol. **161**(4), 793–804 (2003)
35. Lampugnani, M.G., Orsenigo, F., Gagliani, M.C., Tacchetti, C., Dejana, E.: Vascular endothelial cadherin controls. J. Cell Biol. **174**(4), 593–604 (2006)
36. Linehan, W.M., Ricketts, C.J.: Kidney cancer in 2016: RCC - advances in targeted therapeutics and genomics. Nat. Rev. Urol. **14**(2), 76–78 (2016)
37. Lobov, I.B., Renard, R.A., Papadopoulos, N., Gale, N.W., Thurston, G., Yancopoulos, G.D., et al.: Delta-like ligand 4 (dll4) is induced by VEGF as a negative regulator of angiogenic sprouting. Proc. Natl. Acad. Sci. **104**(9), 3219–3224 (2007)
38. Mada, J., Matsuya, K., Yura, F., Kurihara, H., Tokihiro, T.: A mathematical modeling of angiogenesis [in Japanese]. JSIAM **26**, 105–123 (2016)
39. Matsuya, K., Yura, F., Mada, J., Kurihara, H., Tokihiro, T.: A discrete mathematical model for angiogenesis. SIAM J Appl. Math. **76**, 2243–2259 (2016)
40. Maxwell, P.H., Wiesener, M.S., Chang, G.W., Clifford, S.C., Vaux, E.C., Cockman, M.E., et al.: The tumour suppressor protein VHL targets hypoxia-inducible factors for oxygen-dependent proteolysis. Nature **399**(6733), 271–275 (1999)
41. Miller, K., Wang, M., Gralow, J., Dickler, M., Cobleigh, M., Perez, E.A., et al.: Paclitaxel plus bevacizumab versus paclitaxel alone for metastatic breast cancer. N. Engl. J. Med. **357**(26), 2666–2676 (2007)
42. Oda, H., Takeichi, M.: Structural and functional diversity of cadherin at the adherens junction. J. Cell Biol. **193**(7), 1137–1146 (2011)
43. Palachanis, D., Szabó, A., Merks, R.M.: Particle-based simulation of ellipse-shaped particle aggregation as a model for vascular network formation. Comput. Part. Mech. **2**(4), 371–379 (2015). https://doi.org/10.1007/s40571-015-0064-5
44. Ridley, A., Schwartz, M.A., Burridge, K., Firtel, R.A., Ginsberg, M.H., Borisy, G., et al.: Cell migration: integrating signals from front to back. Science **302**(5651), 1704–1710 (2003)
45. Risau, W.: Mechanisms of angiogenesis. Nature **386**(6626), 671–674 (1997)

46. Rorth, P.: Fellow travellers: Emergent properties of collective cell migration. EMBO Rep. **13**(11), 984–991 (2012)
47. Saharinen, P., Eklund, L., Alitalo, K.: Therapeutic targeting of the angiopoietin-tie pathway. Nat. Rev. Drug Discov. **16**(9), 635–661 (2017)
48. Sandler, A., Gray, R., Perry, M.C., Brahmer, J., Schiller, J.H., Dowlati, A., et al.: Paclitaxel–carboplatin alone or with bevacizumab for non–small-cell lung cancer. N. Engl. J. Med. **355**(24), 2542–2550 (2006)
49. Santos-Oliveira, P., Correia, A., Rodrigues, T., Ribeiro-Rodrigues, T.M., Matafome, P., Rodriguez-Manzaneque, J.C., Seiça, R., Girão, H., Travasso, R.D.M.: The force at the tip - modelling tension and proliferation in sprouting angiogenesis. PLOS Comput. Biol. **11**, e1004436 (2015)
50. Scianna, M., Preziosi, L.: Multiscale developments of the cellular Potts model. Multiscale Model. Simul. **10**, 342–382 (2012)
51. Secomb, T.W., Alberding, J.P., Deshirst, M.W., Pries, A.R.: Angiogenesis: an adaptive dynamic biological patterning problem. PLOS Comput. Biol. **9**, e1002983 (2013)
52. Shimamoto, Y., Tamura, S., Masumoto, H., Maeshima, K.: Nucleosome–nucleosome interactions via histone tails and linker DNA regulate nuclear rigidity. Mol. Biol. Cell **28**(11), 1580–1589 (2017). https://doi.org/10.1091/mbc.E16-11-0783
53. Soga, N., Connolly, J.O., Chellaiah, M., Kawamura, J., Hruska, K.A.: Rac regulates vascular endothelial growth factor stimulated motility. Cell Commun. Adhes. **8**(1), 1–13 (2009)
54. Spill, F., Guerrero, P., Alarcon, T., Maini, P.K., Byrne, H.M.: Mesoscopic and continuum modelling of angiogenesis. Math. Biol. **70**, 485–532 (2015)
55. Stéphanou, A., Floc'h, S.L., Chauvière, A.: A hybrid model to test the importance of mechanical cues driving cell migration in angiogenesis. Math. Modell. Nat. Phenom. **10**, 142–165 (2015)
56. Sugihara, K., Nishiyama, K., Fukuhara, S., Uemura, A., Arima, S., Kobayashi, R., Köhn-Luque, A., Mochizuki, N., Suda, T., Ogawa, H., Kurihara, H.: Autonomy and non-autonomy of angiogenic cell movements revealed by experiment-driven mathematical modeling. Cell Rep. **13**, 1814–1827 (2015)
57. Sugihara, K., Nishiyama, K., Fukuhara, S., Uemura, A., Arima, S., Kobayashi, R., et al.: Autonomy and non-autonomy of angiogenic cell movements revealed by experiment-driven mathematical modeling. Cell Rep. **13**(9), 1814–1827 (2015)
58. Suzuki, S., Sano, K., Tanihara, H.: Diversity of the cadherin family: evidence for eight new cadherins in nervous tissue. Cell. Regul. **2**(4), 261–270 (1991)
59. Takeichi, M.: Historical review of the discovery of cadherin, in memory of Tokindo Okada. Develop. Growth Differ. **60**(1), 3–13 (2018)
60. Takubo, N., Yura, F., Naemura, K., Yoshida, R., Tokunaga, T., Tokihiro, T., Kurihara, H.: Cohesive and anisotropic vascular endothelial cell motility driving angiogenic morphogenesis. Sci. Rep. **9**, 9304 (2019)
61. Tanihara, H., Kido, M., Obata, S., Heimark, R.L., Davidson, M., St John, T., et al.: Characterization of cadherin-4 and cadherin-5 reveals new aspects of cadherins. J. Cell Sci. **107**(Pt 6), 1697–1704 (1994)
62. Tong, S., Yuan, F.: Numerical simulations of angiogenesis in the cornea. Microvasc. Res. **61**, 14–27 (2001)
63. Vestweber, D.: VE-cadherin: the major endothelial adhesion molecule controlling cellular junctions and blood vessel formation. Arterioscl. Thromb. Vascul. Biol. **28**(2), 223–232 (2008)
64. Vittet, D., Buchou, T., Schweitzer, A., Dejana, E., Huber, P.: Targeted null-mutation in the vascular endothelial-cadherin gene impairs the organization of vascular-like structures in embryoid bodies. Proc. Natl. Acad. Sci. U. S. A. **94**(12), 6273–6278 (1997)
65. Wu, J., Tang, Y., Liang, X.: Targeting VEGF pathway to normalize the vasculature: an emerging insight in cancer therapy. OncoTargets Therapy **11**, 6901–6909 (2018)
66. Yang, Y., Xie, P., Opatowsky, Y., Schlessinger, J.: Direct contacts between extracellular membrane-proximal domains are required for VEGF receptor activation and cell signaling. Proc. Natl. Acad. Sci. U. S. A. **107**(5), 1906–1911 (2010)

67. Yang, J.C., Haworth, L., Sherry, R.M., Hwu, P., Schwartzentruber, D.J., Topalian, S.L., et al.: A randomized trial of bevacizumab, an anti-vascular endothelial growth factor antibody, for metastatic renal cancer. N. Engl. J. Med. **349**(5), 427–434 (2003)
68. Zhang, L., Bhaloo, S.I., Chen, T., Zhou, B., Xu, Q.: Role of resident stem cells in vessel formation and arteriosclerosis. Circ. Res. **122**(11), 1608–1624 (2018)

Chapter 3
Synchronization and Fluctuation of Cardiac Muscle Cells

Tatsuya Hayashi, Kenji Yasuda, and Guanyu Zhou

3.1 Introduction

Following on from the dramatic progress in the life sciences in the twentieth century, starting with determination of the way in which genetic information is stored, encoded, and transmitted, another challenge has arisen regarding epigenetic information. Epigenetic information is complementary to genetic information and essential to understand the entire landscape of living systems, such as how living cells can choose, reserve, share, and inherit acquired epigenetic information among neighboring cells and between generations. As we move into the post-genomic/proteomic era, such complementarity to genetic information should become more apparent. The cells in a group are individual entities, and differences arise even among cells with identical genetic information that have developed under the same conditions. These cells respond differently to perturbations [42]. Why and how do these differences arise? How are these differences of individual cells ironed out when they become groups, clusters, or tissues? We call this behavior the "community effect" of cells as induced uniformity. To understand the community

T. Hayashi (✉)
Graduate School of Information Science and Technology, Hokkaido University, Sapporo, Hokkaido, Japan
e-mail: thayashi@ist.hokudai.ac.jp

K. Yasuda
Graduate School of Advanced Science and Engineering, Waseda University, Shinjuku-ku, Tokyo, Japan
e-mail: yasuda@waseda.jp

G. Zhou
Institute of Fundamental and Frontier Sciences, University of Electronic Science and Technology of China, Chengdu, China
e-mail: zhoug@uestc.edu.cn

T. Tokihiro (ed.), *Mathematical Modeling for Genes to Collective Cell Dynamics*, Theoretical Biology, https://doi.org/10.1007/978-981-16-7132-6_3

effect, we need to understand the potential underlying differences of cells, and why and how their characteristics change when they form networks as epigenetic information.

If we are to obtain a comprehensive understanding of a living system, we need to analyze its epigenetic information, such as adaptation processes and community effect in a group of cells. As cells are minimal units in terms of both genetic and epigenetic information, we must analyze their epigenetic information starting from the twin complementary perspectives of cell regulation being an "algebraic" system (with emphasis on temporal aspects; nongenetic adaptation) and a "geometric" system (with emphasis on spatial aspects; spatial pattern-dependent community effect) using identified single cells and their patterned groups. We thus commenced a series of studies to analyze the epigenetic information of single cells and the spatial structures of cell networks to expand our understanding of how the fates of living systems are determined and how they can be changed.

The importance of understanding epigenetic information is expected to become apparent in cell-based biological and medical fields such as cell-based drug screening and the regeneration of organs from stem cells, fields in which phenomena cannot be interpreted without taking epigenetic factors into account. We thus started a series of studies focusing on developing a system that could be used to evaluate the epigenetic information in cells by continuously observing specific examples of cells and their interactions under fully controlled conditions as a constructive experimental method. However, the issues of limitations regarding the quality of cells and control of their conditions remained. Mathematical modeling is one of the most powerful approaches to overcome these problems.

In this chapter, a mathematical approach for analyzing the synchronization behavior of spontaneously beating cardiomyocytes was examined, starting from modeling of the firing of cardiomyocytes and progressing to spatially arranged cardiomyocyte networks, based on the twin complementary perspectives of cell regulation, namely, as an "algebraic" system (emphasis on temporal aspects) and as a " geometric" system (emphasis on spatial aspects). Our experimental and mathematical results on the community effect in the synchronization behavior of beating in cardiomyocyte networks are introduced and discussed.

3.2 The Stochastic Phase Models for the Cardiomyocyte Beating

Massive mathematical models have been proposed to investigate the mechanism of cardiomyocyte beating. For example, the work [10] studies an elaborated mathematical model composed of a large number of equations, which looks into the complex electrophysiological processes causing cardiomyocyte synchronization. On the other hand, using just a few ordinary equations, one can reproduce the key phenomenon of the membrane currents and action potentials (see, for example,

[22, 37]), such as the famous Hodgkin–Huxley model, the FitzHugh–Nagumo model, and the Van der Pol model. In this section, we focus on investigating the statistical behavior of beating/synchronization period of cardiomyocyte. To explain the essence of synchronization period, we can regard the cardiac muscle cells as oscillators, to which the phase model is well applicable [29, 32, 56]. It is also regarded as the well-known integrate-and-fire model which has been widely used as a spiking neuron model [2, 23, 40]. However, to capture the features of cardiomyocyte beating, we have to incorporate the conventional stochastic phase models with three important conceptions: irreversible at firing, a refractory period after firing, and induced pulsation associated with firing of neighboring cells. In this section, we introduce the stochastic phase models for the beating of the isolated and coupled cardiac muscle cells. The theoretical analysis concerned with the synchronization period is provided, which involves the stochastic equation, the Itô formula, and the calculation of expected value, variance, and coefficient of variance (CV) of the beating/synchronization period.

3.2.1 Some Preliminaries for the Stochastic Phase Model

Before describing the stochastic phase models for the cardiomyocyte beating, we briefly introduce some mathematical preliminaries on the phase model and stochastic differential equation.

3.2.1.1 The Phase Model

We start from a simple phase model. Let ϕ be the phase of an oscillator with intrinsic frequency (or drift) $\mu > 0$ and initial state $\phi(0) = 0$. The phase model is given by

$$\phi(t) = \mu t. \tag{3.1}$$

Assuming that the phase returns to 0 when approaching 2π, we see that $T = \frac{2\pi}{\mu}$ is the period of the oscillator. We can also write (3.1) into an equivalent differential form:

$$d\phi(t) = \mu dt, \tag{3.2a}$$

$$\phi(0) = 0. \tag{3.2b}$$

Think of the oscillator as a cardiomyocyte, which beats when the phase reaches 2π and then returns to 0 immediately to begin a new beating process (see Fig. 3.1a). The phase equation (3.1) describes the rhythmic beating with period T.

In general, one can consider the phase model with time-dependent and state-dependent drift, that is, $\mu(t, \phi(t))$ is a function depending on t and ϕ. Then, the

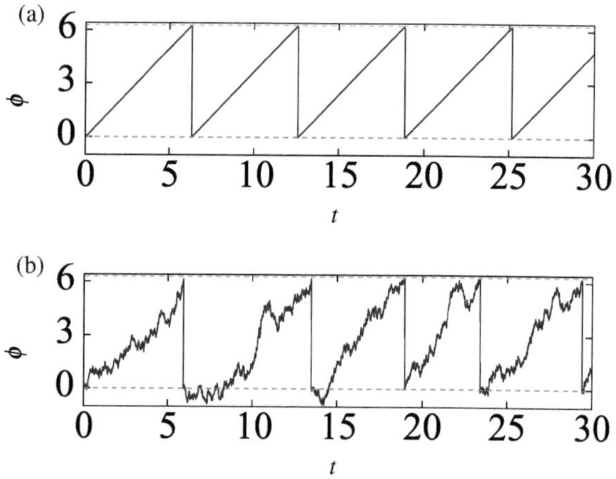

Fig. 3.1 (**a**) The phase model with constant drift $\mu = 1$, where the beating interval is 2π. (**b**) The phase model with drift $\mu = 1$ and "white noise" ζ, where the beating interval varies each time

phase model with initial value ϕ_0 becomes

$$d\phi = \mu(t, \phi)dt, \quad \phi(0) = \phi_0.$$

The above equation is equivalent to the following integration form:

$$\phi(t) = \int_0^t \mu(s, \phi(s))\, ds + \phi_0, \tag{3.3}$$

which is also called the "integrate-and-fire" model.

However, for a cardiomyocyte, the beating process is often affected by the internal/external noise. As a result, the beating interval varies each time (see Fig. 3.1b for an example of beating process with noise).

3.2.1.2 The Brownian Motion and White Noise

Incorporating the phase model (3.2) with noise effect, we write the phase model in a formal way:

$$d\phi(t) = \mu dt + \sigma\zeta(t), \tag{3.4a}$$

$$\phi(0) = 0, \tag{3.4b}$$

where $\zeta(t)$ denotes the "white noise" (which has been widely applied in many mathematical models), and σ is a constant representing the strength of the noise. In general, one can take σ as a function of ϕ, i.e., $\sigma(\phi)$. Since the white noise can be regarded as the time derivative of Brownian motion (or called the Winner process) denoted by $W(t)$, (3.4a) becomes

$$d\phi(t) = \mu dt + \sigma dW(t), \tag{3.5}$$

which is a stochastic differential equation. Figure 3.1b shows a sample path of ϕ with $\mu = \sigma = 1$.

Before we discuss the stochastic differential equation (3.5), let us explain the conceptions of the Brownian motion and white noise, where we need the Gaussian (or normal) distribution.

For $\mu \geq 0$ and $\sigma > 0$, $N(\mu, \sigma^2)$ represents the Gaussian (or normal) distribution with mean μ and variance σ^2. The probability density function of $N(\mu, \sigma^2)$ is given by

$$f(x) = \frac{1}{\sqrt{2\pi\sigma^2}} e^{-\frac{|x-\mu|^2}{2\sigma^2}}, \quad -\infty < x < \infty. \tag{3.6}$$

For an $N(\mu, \sigma^2)$ random variable X (i.e., $X \sim N(\mu, \sigma^2)$), one can compute the probability of the event $a < X < b$ using the density function f:

$$P(a < X < b) = \int_a^b f(x)\, dx.$$

The expected value (or mean) of X is calculated as

$$\begin{aligned} \mathbf{E}(X) &= \int_{-\infty}^{\infty} x dP(X \leq x) = \int_{-\infty}^{\infty} x f(x) dx \\ &= \int_{-\infty}^{\infty} x \frac{1}{\sqrt{2\pi\sigma^2}} e^{-\frac{|x-\mu|^2}{2\sigma^2}} dx = \mu. \end{aligned} \tag{3.7}$$

And we derive the variance of X:

$$\begin{aligned} \mathbf{Var}(X) &= \mathbf{E}([X - \mathbf{E}(X)]^2) \\ &= \mathbf{E}(X^2) - [\mathbf{E}(X)]^2 \\ &= \int_{-\infty}^{\infty} x^2 \frac{1}{\sqrt{2\pi\sigma^2}} e^{-\frac{|x-\mu|^2}{2\sigma^2}} dx - \mu^2 = \sigma^2. \end{aligned} \tag{3.8}$$

The two (Gaussian) random variables X_1 and X_2 are independent means the probability of the events $a_1 < X_1 < b_1$ and $a_2 < X_2 < b_2$ equals the product of $P(a_1 < X_1 < b_1)$ and $P(a_2 < X_2 < b_2)$, i.e.,

$$P(a_1 < X_1 < b_1, a_2 < X_2 < b_2) = P(a_1 < X_1 < b_1) P(a_2 < X_2 < b_2),$$

where $\{a_i\}_{i=1}^2$ and $\{b_i\}_{i=1}^2$ are any real numbers with $a_i < b_i$.

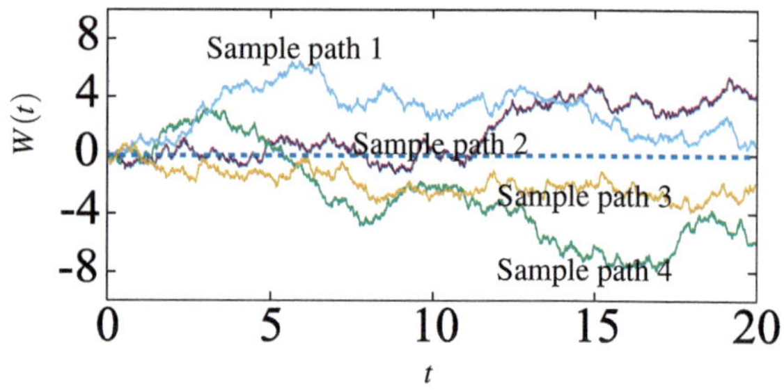

Fig. 3.2 Four sample paths of Brownian motion $W(t)$

For two independent random variables X_1 and X_2, we have

$$\mathbf{E}(X_1X_2) = \mathbf{E}(X_1)\mathbf{E}(X_2). \tag{3.9}$$

With the help of Gaussian random variable, we introduce the one-dimensional Brownian motion.

A collection of random variables $\{X_t \mid t \geq 0\}$ is called a stochastic process. The Brownian motion $\{W(t) \mid t \geq 0\}$ is a stochastic process satisfying:

- $W(0) = 0$.
- For all $0 \leq s < t$, $W(t) - W(s) \sim N(0, t - s)$.
- For any $0 < t_1 < t_2 < \cdots < t_n$, the random variables $W(t_1)$, $W(t_2) - W(t_1), \cdots, W(t_n) - W(t_{n-1})$ are independent.

We plot four sample paths (or trajectories) of Brownian motion $W(t)$ (see Fig. 3.2). Noting that $W(t) \sim N(0, t)$, we have the mean and variance of $W(t)$:

$$\mathbf{E}[W(t)] = 0, \quad \mathbf{Var}[W(t)] = \mathbf{E}[W^2(t)] - 0^2 = t. \tag{3.10}$$

Moreover, for $t > s \geq 0$,

$$\begin{aligned} \mathbf{E}[W(t)W(s)] &= \mathbf{E}[(W(t) - W(s) + W(s))W(s)] \\ &= \mathbf{E}[(W(t) - W(s))W(s)] + \mathbf{E}[W^2(s)] \\ &= \mathbf{E}[W(t) - W(s)]\mathbf{E}[W(s)] + s \qquad \text{(by (3.9))} \\ &= 0 + s, \end{aligned} \tag{3.11}$$

where we have used the fact that $W(t) - W(s) \sim N(0, t - s)$, $W(s) \sim N(0, s)$, and the random variables $W(t) - W(s)$ and $W(s)$ are independent.

We have described Brownian motion. As mentioned above, the "white noise" $\zeta(t)$ is the formal time derivative of $W(t)$, i.e., $\zeta(t) = \frac{dW(t)}{dt}$. However, in fact, the sample path of $W(t)$ is not differentiable for $t \geq 0$; that is, $\frac{dW(t)}{dt}$ does not really exist (in the classical sense).

To get a better understanding of the "white noise" (or $dW(t)$), we introduce Itô's integral to interpret $dW(t)$ in the integral form.

3.2.1.3 Itô's Integral

Analogously to the integral form (3.3) of the phase model, we write the model (3.5) into the integral equation

$$\phi(t) = \int_0^t \mu dt + \int_0^t \sigma dW(s) + \phi_0. \tag{3.12}$$

The task is to provide a proper definition of the integral that involves $dW(s)$, i.e., $\int_0^t \sigma \, dW(s)$, such that (3.12) makes sense. For the simple case that μ and σ are constants, we immediately have (by $\int_0^t dW(s) = W(t)$)

$$\phi(t) = \mu t + \sigma W(t) + \phi_0.$$

See Fig. 3.1b for a sample path of $\phi(t)$ with $\mu = \sigma = 1$ and $\phi_0 = 0$ and under the setting that ϕ jumps to 0 when $\phi(t)$ reaching 2π. However, for a stochastic process $X(t)$, the definition of $\int_0^T X(t) \, dW(t)$ is not straightforward. The famous Itô's integral addresses this issue.

The rigorous mathematical definition of Itô's integral involves the filtration and measurability theories of stochastic process, which is omitted in the following argument and replaced by some intuitive description. One can refer to [6, 35] for a detailed mathematical definition.

First, we consider a simple case that $X(t)$ is a step process: for $0 = t_0 < t_1 < \cdots < t_n = T$,

$$X(t) = X_k \quad \text{for } t_k \le t < t_{k+1}, \quad k = 0, 1, \ldots, n-1,$$

where $\{X_k\}_{k=0}^n$ are the random variables independent of t, satisfying:

(c1) X_k is independent of the information of $W(t)$ for all $t \ge t_k$.
(c2) $\mathbf{E}(\int_0^T |X(t)|^2 \, dt) < \infty$.

Then, the Itô stochastic integral of $X(t)$ on $(0, T)$ is defined by

$$\int_0^T X(t) \, dW(t) = \sum_{k=0}^{n-1} X_k (W(t_{k+1}) - W(t_k)). \tag{3.13}$$

It is apparent that, for any constants a and b and the step processes X_1 and X_2 satisfying (c1) and (c2), the following equality holds:

$$\int_0^T aX_1(t) + bX_2(t)\, dW(t) = a \int_0^T X_1\, dW(t) + b \int_0^T X_2\, dW(t). \tag{3.14}$$

In statistical analysis, the expected value (mean) and variance (fluctuation) are important features of a random variable. Let us investigate the expected value involving Itô's integral.

The assumption that $X(t)$ only depends on the past history of the Brownian motion $\{W(s) \mid s < t\}$ but is independent of the future behavior $\{W(s) \mid s \geq t\}$ plays crucial role in the obtention of the following properties of Itô's integral.

According to the definition (3.13),

$$\mathbf{E}\left(\int_0^T X(t)\, dW(t)\right) = \sum_{k=0}^{n-1} \mathbf{E}\left[X_k(W(t_{k+1}) - W(t_k))\right].$$

The assumption (c1) implies that X_k is independent of $W(t_{k+1})$ and $W(t_k))$. As a result,

$$\begin{aligned}\mathbf{E}\left[X_k(W(t_{k+1}) - W(t_k))\right] &= \mathbf{E}(X_k)\mathbf{E}(W(t_{k+1}) - W(t_k)) \\ &= 0 \quad (\text{by } W(t_{k+1}) - W(t_k) \sim N(0, t_{k+1} - t_k)).\end{aligned}$$

Hence, we conclude

$$\mathbf{E}\left(\int_0^T X(t)\, dW(t)\right) = 0. \tag{3.15}$$

Next, we consider the expected value of $\left(\int_0^T X(t)\, dW(t)\right)^2$. By the definition (3.13),

$$\begin{aligned}\mathbf{E}\left[\left(\int_0^T X(t)\, dW(t)\right)^2\right] = \sum_{k=0}^{n-1}\sum_{j=0}^{n-1} \mathbf{E}\big[X_k(W(t_{k+1}) \\ - W(t_k))X_j(W(t_{j+1}) - W(t_j))\big].\end{aligned}$$

For $k \neq j$, without loss of generality, we assume $k > j$. Since $(W(t_{k+1}) - W(t_k))$ and $(W(t_{j+1}) - W(t_j))$ are independent, and X_j is only dependent upon the past information $\{W(t) \mid t < t_j\}$ and independent of the future behavior $\{W(t) \mid t \geq t_j\}$, we find that $(W(t_{k+1}) - W(t_k))$ is independent of $X_k X_j (W(t_{j+1}) - W(t_j))$, which implies

$$\begin{aligned}
&\mathbf{E}\left[X_k(W(t_{k+1}) - W(t_k))X_j(W(t_{j+1}) - W(t_j))\right] \\
&\quad = \mathbf{E}(W(t_{k+1}) - W(t_k))\mathbf{E}[X_j X_k(W(t_{j+1}) - W(t_j))] \\
&\quad = 0 \quad (\text{because } W(t_{k+1}) - W(t_k) \sim N(0, t_{k+1} - t_k)).
\end{aligned}$$

Therefore,

$$\mathbf{E}\left[\left(\int_0^T X(t)\, dW(t)\right)^2\right] = \sum_{k=0}^{n-1} \mathbf{E}\left[X_k^2(W(t_{k+1}) - W(t_k))^2\right].$$

Again, the assumption (c1) yields the independence between X_k^2 and $(W(t_{k+1}) - W(t_k))^2$, which gives

$$\begin{aligned}
\mathbf{E}\left[X_k^2(W(t_{k+1}) - W(t_k))^2\right] &= \mathbf{E}(X_k^2)\mathbf{E}\left[(W(t_{k+1}) - W(t_k))^2\right] \\
&= \mathbf{E}(X_k^2)\mathbf{E}\left[W^2(t_{k+1}) - 2W(t_{k+1})W(t_k) + W^2(t_k)\right] \\
&= \mathbf{E}(X_k^2)(t_{k+1} - 2t_k + t_k) \quad (\text{by (3.10) and (3.11)}) \\
&= \mathbf{E}(X_k^2)(t_{k+1} - t_k) = \mathbf{E}\left[\int_{t_k}^{t_{k+1}} X^2(t)\, dt\right],
\end{aligned}$$

where we have used the fact that $X(t) = X_k$ for $t_k \le t < t_{t_{k+1}}$. Summing up the above equation from $k = 0$ to $k = n - 1$ yields

$$\mathbf{E}\left[\left(\int_0^T X(t)\, dW(t)\right)^2\right] = \mathbf{E}\left(\int_0^T X^2(t)\, dt\right). \tag{3.16}$$

We have introduced Itô's integral for the step process $X(t)$ and derived the properties (3.14), (3.15), and (3.16). For arbitrary progressive measurable process $X(t)$ (generally speaking, X is a progressive measurable process means that $X(t)$ is independent of the future behavior of the Brownian motion $\{W(s) \mid s \ge t\}$ and is well defined and measurable providing the past information of the Brownian motion $\{W(s) \mid s < t\}$) satisfying the boundedness $\int_0^T |X(t)|^2\, dt < \infty$, we can approximate $X(t)$ by a sequence of step processes $\{X^{(m)}(t)\}_{m=1}^{\infty}$ and define Itô's integral as the limitation of $\int_0^T X^{(m)}(t)\, dW(t)$, i.e.,

$$\int_0^T X(t)\, dW(t) = \lim_{m\to\infty} \int_0^T X^{(m)}(t)\, dW(t).$$

The readers can refer to [6, 35] for the detailed approximation technique and convergence analysis. In fact, one can verify that the properties (3.14), (3.15), and (3.16) are also satisfied. Therefore, we have extended Itô's integral to arbitrary progressive measurable process $X(t)$ with bounded L^2-norm (i.e., $\int_0^T |X(t)|^2\, dt < \infty$).

Itô's integral ensures the meaning of the integral equation (3.12). Usually, it is more convenient to write (3.12) into the differential form (3.5), where the white noise $\zeta = dW(t)$ should be understood in the sense of Itô's integral.

3.2.1.4 Itô's Formula

Given a smooth function $g(x)$ and a stochastic process $X(t)$ satisfying the following stochastic differential equation:

$$dX(t) = \mu(X(t))dt + \sigma(X(t))dW(t), \tag{3.17a}$$

$$X(0) = X_0, \tag{3.17b}$$

where $\mu(X(t))$ and $\sigma(X(t))$ are two functions depending on $X(t)$, we aim to investigate the random variable $g(X(t))$.

We give an explicit statement of Itô's formula. Assume that $g(x)$ is a twice continuously differentiable function and X satisfies (3.17). Then, we have

$$g(X(t)) - g(X(0)) = \int_0^t g'\mu + \frac{1}{2}g''\sigma^2\, ds + \int_0^t g'\sigma\, dW(s), \tag{3.18}$$

where g', g'', σ, and μ are all functions of $X(s)$. One can write the above formula in a more compact form:

$$\begin{aligned} dg(X) &= g'(X)dX + \frac{1}{2}g''(X)\sigma^2 dt \\ &= \left(g'(X)\mu + \frac{1}{2}g''(X)\sigma^2\right) dt + g'(X)\sigma dW. \end{aligned} \tag{3.19}$$

The readers can refer to [6, 35] for the proof of (3.18).

We introduce another expression of (3.19) by the following expansion of $dg(X)$:

$$dg(X) = g'(X)dX + \frac{1}{2}g''(X)(dX)^2, \tag{3.20}$$

where $(dX)^2$ is decomposed by

$$(dX)^2 = \mu^2(dt)^2 + \mu\sigma(dtdW + dWdt) + \sigma^2(dW)^2.$$

Roughly speaking, we have

$$(dW)^2 = dt \quad \left(\text{or equivalently, } dW = (dt)^{\frac{1}{2}}\right),$$

according to which $dWdt = dtdW = (dt)^{\frac{3}{2}}$. And we omit the terms of $(dt)^2$, $dWdt$, and $dtdW$ since they are of a higher order than dt. Then, the expansion (3.20) reduces to (3.19). But rigorously speaking, dX and dW have no meaning alone.

If we apply the ordinary differential chain rule, then we get

$$dg(X) = g'(X)dX = g'(X)\mu dt + g'(X)\sigma dW,$$

which is not correct.

For $g(x, t)$ with continuous differential $\frac{\partial g}{\partial t}$, $\frac{\partial g}{\partial x}$, and $\frac{\partial^2 g}{\partial x^2}$, Itô's formula becomes

$$g(X(t), t) - g(X(0), 0) = \int_0^t \frac{\partial g}{\partial s} + \frac{\partial g}{\partial x}\mu + \frac{1}{2}\frac{\partial^2 g}{\partial x^2}\sigma^2 \, ds + \int_0^t \frac{\partial g}{\partial x}\sigma \, dW(s), \tag{3.21}$$

or equivalently,

$$\begin{aligned} dg(X, t) &= \frac{\partial g}{\partial t}dt + \frac{\partial g}{\partial x}dX + \frac{1}{2}\frac{\partial^2 g}{\partial x^2}\sigma^2 dt \\ &= \left(\frac{\partial g}{\partial t}dt + \frac{\partial g}{\partial x}\mu + \frac{1}{2}\frac{\partial^2 g}{\partial x^2}\sigma^2\right) dt + \frac{\partial g}{\partial x}\sigma dW. \end{aligned} \tag{3.22}$$

We end up this section with an application of Itô's formula. Consider a particle moves with a constant drift μ in one-dimensional space starting from the position $x = 0$. Moreover, the particle is affected by a white noise with strength σ. There are two barriers at $x = \pm 1$, and we assume that the particle is absorbed when touching the barriers. We denote by $\phi(t)$ the position of the particle at time t. Suppose that the particle touches the barriers at time τ, i.e., $\phi(\tau) = \pm 1$. τ is a random variable. What is the expected value (mean) of τ?

To compute the expected value of τ, we first notice that ϕ satisfies the stochastic differential equation

$$d\phi(t) = \mu dt + \sigma dW(t), \quad \phi(0) = 0.$$

See Fig. 3.3 for two examples. Noting that

$$\phi(t) = \mu t + \sigma W(t),$$

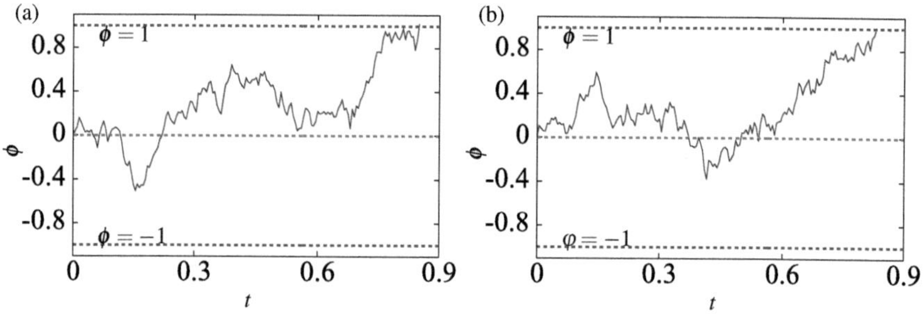

Fig. 3.3 (**a**) A sample path of $\phi(t)$ with $\mu = 0$ and $\sigma = 1$. (**b**) A sample path of $\phi(t)$ with $\mu = 1$ and $\sigma = 1$. The movement is stopped when reaching the boundary ± 1

ϕ is the stochastic process with mean

$$\mathbf{E}(\phi(t)) = \mu t$$

and variance

$$\mathbf{Var}(\phi(t)) = \mathbf{E}(\phi^2(t)) - [\mathbf{E}(\phi(t))]^2 = \sigma^2 t.$$

In fact, $\phi(t)$ is called the (μ, σ) Brownian motion, and $W(t)$ is the standard Brownian motion. And we see that $\phi(t) \sim N(\mu t, \sigma^2 t)$ and $W(t) \sim N(0, t)$.

By Itô's formula (3.18), for a function g with continuous g' and g'',

$$g(\phi(\tau)) - g(0) = \int_0^\tau g'\mu + \frac{1}{2}g''\sigma^2 \, dt + \int_0^\tau g'\sigma \, dW(t).$$

Since τ is regarded as the first time that the particle reaches $x = \pm 1$, we see that $g(\phi(\tau)) = g(\pm 1)$. Assume that g is the solution of the following boundary value problem:

$$g'\mu + \frac{1}{2}g''\sigma^2 = -1 \qquad \text{for } -1 < x < 1, \tag{3.23a}$$

$$g(\pm 1) = 0. \tag{3.23b}$$

Then, we find that

$$0 - g(0) = -\int_0^\tau 1 \, dt + \int_0^\tau g'\sigma \, dW(t) = -\tau + \int_0^\tau g'\sigma \, dW(t).$$

Applying (3.15),

$$\begin{aligned} 0 - g(0) &= -\mathbf{E}(\tau) + \mathbf{E}\left(\int_0^\tau g'\sigma \, dW(t)\right) \\ &= -\mathbf{E}(\tau) + 0. \end{aligned}$$

Therefore, the expected value of the absorbed time is given by

$$\mathbf{E}(\tau) = g(0).$$

The problem reduces to solving the boundary value problem (3.23). For the case that $\mu = 0$, it is obvious that

$$g(x) = \sigma^{-2}(1 - x^2),$$

which yields $\mathbf{E}(\tau) = \sigma^{-2}$.

We have introduced some preliminaries of stochastic differential equation. Next, we turn attention to our stochastic phase models for the beating process of cardiomyocyte, where we apply the properties of Itô's integral and Itô's formula for analysis.

3.2.2 *The Phase Model for an Isolated Cardiomyocyte*

The beating process for an isolated cardiomyocyte is regarded as the increase of a stochastic phase function from 0 to 2π, where the phase starts from 0, increases with an intrinsic frequency μ, and is effected by a white noise with strength σ. When the phase approaches 2π, we say the cell beats and the phase then returns to 0. Hence, from 0 to 2π, the cell completes an oscillation cycle (see Fig. 3.4). To incorporate the irreversibility of beating, we impose a reflective boundary at 0 state (see Figs. 3.4b and 3.5a).

Let $\phi(t)$ denote the phase of an isolated cardiomyocyte at time t. The model is stated as follows:

$$d\phi(t) = \mu dt + \sigma dW(t) + dL(t), \tag{3.24}$$

where μ denotes the intrinsic frequency, σ represents the noise strength, and $W(t)$ stands for the standard Brownian motion. $dW(t)$ is the generalized derivative of $W(t)$, which is known as the Gaussian white noise. We impose the conception of irreversibility after beating. When the cell beats, we have $\phi(t-) = 2\pi$ and $\phi(t) = 0$. Since the phase is affected by noise σdW, when $\phi(t-) = 0$ and $\sigma dW(t-) < 0$, the phase may become negative, i.e., $\phi(t)$ goes back to 2π (see Fig. 3.4b). The irreversibility says that when $\phi(t-) = 0$, ϕ cannot be driven back to 2π by negative

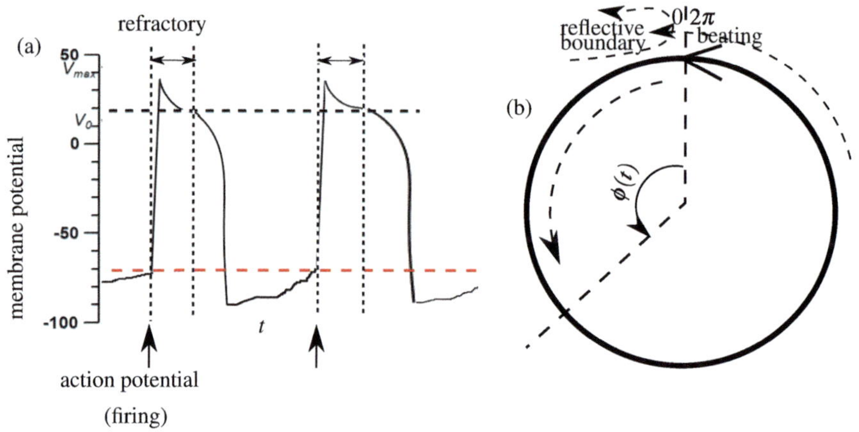

Fig. 3.4 (**a**) The cardiac action potential. (**b**) The beating process is modeled by the increase of the phase function $\phi(t)$ from 0 to 2π. The cell beats when the phase ϕ approaches 2π and then returns to 0 to start a new oscillation circle. The reflective boundary is imposed at $\phi(t) = 0$, which ensures that $\phi \geq 0$ always holds even when $\phi(t) = 0$ and the noise $\sigma dW(t)$ is negative

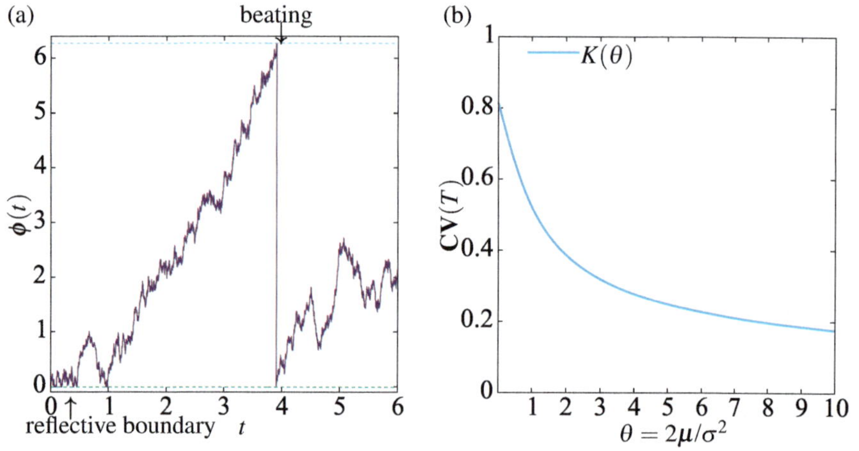

Fig. 3.5 (**a**) A trajectory of $\phi(t)$ with $(\mu, \sigma) = (1, 2)$, $0 \leq t \leq 6$. When ϕ approaches 2π, it returns to 0 immediately. (**b**) The CV of beating interval

noise. To prevent the reversibility of beating, we add the process $L(t)$ to cancel the negative part of noise such that $\phi(t) \geq 0$ always holds (see Fig. 3.5a). Hence, the reflective boundary is described by $L(t)$, which satisfies [9, 41]:

1. $L(t)$ increases only when $\phi(t) = 0$ such that $\phi(t) \geq 0$.
2. $L(t) = 0$ when the cell beats.

Every time ϕ approaches 2π, ϕ returns to 0, which means that ϕ is a renewal process. The beating interval T is a random variable owing to the noise $dW(t)$. Next, we investigate the expected value and variance of T, i.e., the mean and fluctuation of beating interval of cardiomyocyte.

For any function g with continuous differential $\frac{dg}{dx}$ and $\frac{d^2g}{dx^2}$, Itô formula gives

$$g(\phi(T)) - g(\phi(0)) = \int_0^T \left[\mu \frac{dg}{dx}(\phi(t)) + \frac{\sigma^2}{2}\frac{d^2g}{dx^2}(\phi(t))\right] dt + \int_0^T \frac{dg}{dx}(\phi(t))[\sigma dW(t) + dL(t)]. \tag{3.25}$$

By the properties of Itô's integral [6, 35], we have, for any $G(t)$ and $H(t)$ satisfying $\int_0^T |G|^2 \, dt < \infty$ and $\int_0^T |H|^2 \, dt < \infty$,

$$\mathbf{E}\left(\int_0^T G(t)\, dW(t)\right) = 0, \tag{3.26a}$$

$$\mathbf{E}\left(\int_0^T G(t)\, dW(t) \int_0^T H(t)\, dW(t)\right) = \mathbf{E}\left(\int_0^T GH\, dt\right). \tag{3.26b}$$

Equation (3.26a) means that the expected value of Itô's integral with respect to dW equals 0.

Now, in view of $\phi(0) = 0$ and $\phi(T) = 2\pi$, if g satisfies

$$g(2\pi) = 0, \quad \mu\frac{dg}{dx} + \frac{\sigma^2}{2}\frac{d^2g}{dx^2} = -1, \tag{3.27}$$

then it follows from (3.25) that

$$0 - g(0) = -T + \int_0^T \frac{dg}{dx}(\phi(t))\sigma\, dW(t) + \int_0^T \frac{dg}{dx}(\phi(t))\, dL(t).$$

Since $L(t)$ is a process that increases only when $\phi(t) = 0$, $dL(t)$ is nonzero only when $\phi(t) = 0$. Thus, we have

$$\int_0^T \frac{dg}{dx}(\phi(t))\, dL(t) = \int_{\{t|0<t<T,\ \phi(t)=0\}} \frac{dg}{dx}(0)\, dL(t).$$

Therefore, if g satisfies additionally

$$\frac{dg}{dx}(0) = 0, \tag{3.28}$$

then we get

$$0 - g(0) = -T + \int_0^T \frac{dg}{dx}(\phi(t))\sigma\, dW(t) + 0. \tag{3.29}$$

Taking the expected value of the above equation and according to (3.26a),

$$\mathbf{E}[T] = g(0).$$

Now, what is left is to find the function g satisfying (3.27) and (3.28). In fact, one can validate that

$$g(x) = \begin{cases} \dfrac{4\pi^2 - x^2}{\sigma^2} & \text{for } \mu = 0, \\[2ex] \dfrac{2\pi - x}{\mu} - \dfrac{\sigma^2\left(e^{-\frac{2\mu x}{\sigma^2}} - e^{-\frac{4\pi\mu}{\sigma^2}}\right)}{2\mu^2} & \text{for } \mu > 0. \end{cases}$$

As a result, we obtain the mean value of the beating interval T:

$$\mathbf{E}(T) = g(0) = \begin{cases} \dfrac{4\pi^2}{\sigma^2} & \text{for } \mu = 0, \\[2ex] \dfrac{2\pi}{\mu} - \dfrac{\sigma^2\left(1 - e^{-4\pi\mu/\sigma^2}\right)}{2\mu^2} & \text{for } \mu > 0. \end{cases} \tag{3.30}$$

Next, we consider the fluctuation of beating interval T, i.e., the variance $\mathbf{Var}(T) = \mathbf{E}(T^2) - [\mathbf{E}(T)]^2$. By virtue of (3.29),

$$T^2 = g^2(0) + 2g(0)\int_0^T \frac{dg}{dx}(\phi(t))\, dt + \left[\int_0^T \frac{dg}{dx}(\phi(t))\, dt\right]^2.$$

Taking the expectation of the above equation and with the help of (3.26), we derive

$$\mathbf{E}[T^2] = g^2(0) + \sigma^2\mathbf{E}\left[\int_0^T \left|\frac{dg}{dx}(\phi(t))\right|^2 dt\right].$$

Recalling that $g(0) = \mathbf{E}(T)$,

$$\begin{aligned}\mathbf{Var}(T) =& \mathbf{E}\left[\int_0^T \sigma^2\left|\frac{dg}{dx}(\phi(t))\right|^2 dt\right] = \int_0^{2\pi} \sigma^2\left|\frac{dg}{dx}(x)\right|^2 \mathbf{E}\left[\int_0^T 1_{dx}(\phi(t))\, dt\right] \\ =& \mathbf{E}(T)\int_0^{2\pi} \sigma^2\left|\frac{dg}{dx}\right|^2 p(x)\, dx,\end{aligned}$$

where $p(x)dx$ represents that the probability of $\phi(t)$ exists in $[x, x+dx)$ for $t \to \infty$ (see [4, Chapter 9 (1.22) (2.25)]). Via a similar calculation to [9], one can obtain the probability density p:

$$p(x) = \begin{cases} \dfrac{2\pi - x}{2\pi^2} & \text{for } \mu = 0, \\ \dfrac{\theta(e^{2\pi\theta} - e^{\theta x})}{1 + 2\pi\theta e^{2\pi\theta} - e^{2\pi\theta}} & \text{for } \mu > 0, \end{cases}$$

where $\theta = 2\mu/\sigma^2$. With the help of g and p, we calculate the variance of beating interval:

$$\mathbf{Var}(T) = \begin{cases} \dfrac{32\pi^4}{3\sigma^4} & \text{for } \mu = 0, \\ \dfrac{-5 + e^{-4\pi\theta} + 4e^{-2\pi\theta} + 8\pi\theta e^{-2\pi\theta} + 4\pi\theta}{\mu^2\theta^2} & \text{for } \mu > 0. \end{cases} \tag{3.31}$$

The coefficient of variance (CV) of beating interval T is given by

$$\mathbf{CV}(T) = \frac{\sqrt{\mathbf{Var}(T)}}{\mathbf{E}(T)} = \begin{cases} \sqrt{2/3} & \text{for } \mu = 0, \\ K(\theta) & \text{for } \mu > 0, \end{cases} \tag{3.32}$$

where $K(\theta) = \frac{e^{2\pi\theta}\sqrt{-5+e^{-4\pi\theta}+4e^{-2\pi\theta}+8\pi\theta e^{-2\pi\theta}+4\pi\theta}}{1+2\pi\theta e^{2\pi\theta}-e^{2\pi\theta}}$. We find that CV only depends on $\theta = 2\mu/\sigma^2$. Moreover, one can validate that the CV decreases as θ increases, and it has an upper bound $\sqrt{2/3} \approx 0.8165$ (see Fig. 3.5b), i.e.,

$$\mathbf{CV}(T) \uparrow \sqrt{3/2} \qquad \theta \downarrow 0.$$

Hence, the phase model (3.24) is only suitable for the cardiomyocyte with CV less than $\sqrt{2/3}$.

In experiment, the distribution of the beating interval T of each isolated cardiomyocyte has been recorded [28, Figure 3. a, Table 1 and Table 2]). Using the above formulas (3.30) and (3.32), we can determine the parameters (μ, σ) of the phase model (3.24) for each cardiomyocyte from the mean and CV of beating interval obtained from experiments. To validate the efficiency of our model, we compare the numerical simulation results with the experimental data (see Sect. 3.3).

3.2.3 The Phase Model for Two Coupled Cardiomyocytes

For two coupled cardiomyocytes, denoting by ϕ_i the phase function of cell i ($i = 1, 2$), we introduce the reaction term $A_{i,j} f(\phi_j - \phi_i)$ between cell i and cell j, where

$\{A_{i,j}\}_{i,j}$ are constants and f is a 2π-periodic function with $f(x) = -f(-x)$ (for example, $f(x) = \sin(x)$). We denote by (μ_i, σ_i) the intrinsic frequency and noise strength of cell i. Then, (ϕ_1, ϕ_2) satisfies, for $i, j = 1, 2, i \neq j$,

$$d\phi_i(t) = \mu_i dt + A_{i,j} f(\phi_j - \phi_i)dt + \sigma_i dW_i(t) + dL_i(t), \tag{3.33a}$$

$$\phi_i(0) = 0, \tag{3.33b}$$

where dW_i represents the white noise of cell i (W_1 and W_2 are independent Brownian motion), and $L_i(t)$ is the process to cancel the negative noise when $\phi_i(t) = 0$ such that $\phi_i(t) \geq 0$, which ensures the irreversibility of beating (see (i) and (ii) of Sect. 3.2.2).

For two coupled cardiomyocytes, we also incorporate the conception of induced beating and refractory:

(R1) When cardiac muscle cell i beats, if cell j is out of refractory, then both the cells beat.
(R2) When cardiac muscle cell i beats, if cell j is in refractory, then only cell i beats.

In phase model, the beating of cell i is described by the phase ϕ_i approaching 2π. For cell j, we set a refractory threshold $0 \leq B_j < 2\pi$. Then, the above (R1) and (R2) are equivalent to the following statements:

(R1) If $\phi_i(t-) = 2\pi$ and $\phi_j(t-) > B_j$, then both the cells beat and return to 0 phase, that is, $\phi_i(t) = \phi_j(t) = 0$ (see Fig. 3.6a).
(R2) If $\phi_i(t-) = 2\pi$ and $\phi_j(t-) \leq B_j$, then only cell i beats and returns to 0 phase, that is, $\phi_i(t) = 0$, $\phi_j(t) = \phi_j(t-)$ (see Fig. 3.6b).

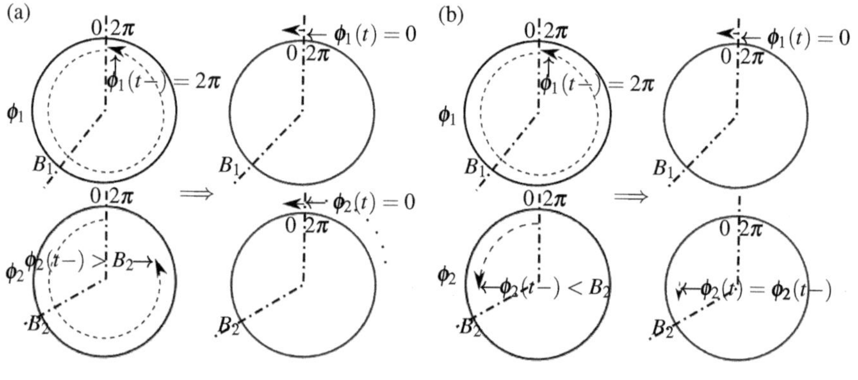

Fig. 3.6 (**a**) (R1) Cell 1 beats and cell 2 is out of refractory, and then cell 2 is induced to beat. $\phi(t-) = 2\pi$ and $\phi_2(t-) > B_2$, then $\phi_1(t) = \phi_2(t) = 0$. (**b**) (R2) Cell 1 beats, but cell 2 is in refractory, and then only cell 1 returns to 0 phase. $\phi(t-) = 2\pi$ and $\phi_2(t-) <= B_2$, then $\phi_1(t) = 0$ and $\phi_2(t) = \phi_2(t-)$

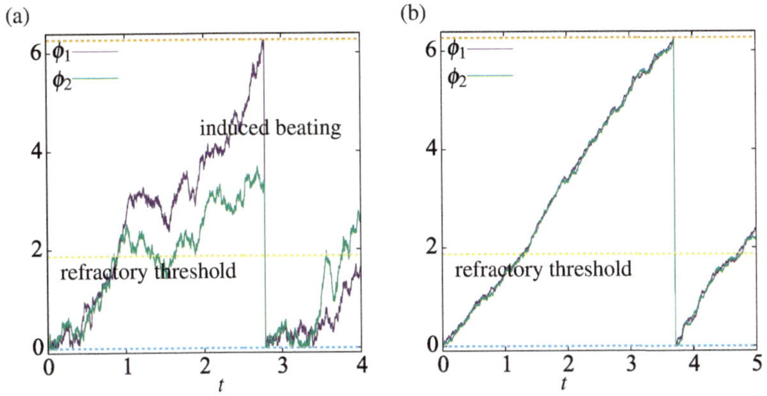

Fig. 3.7 The trajectories of $(\phi_1(t), \phi_2(t))$ with parameters $(\mu_1, \mu_2) = (2, 1)$, $A_{1,2} = A_{2,1} = 2$, and $B_1 = B_2 = 0.6\pi$. **(a)** The strong noise strength $(\sigma_1, \sigma_2) = (1, 1)$. **(b)** The weak noise strength $(\sigma_1, \sigma_2) = (0.3, 0.3)$

In fact, (R1) represents the synchronization of beating, and (R2) is the independent beating of cell i. In Fig. 3.7, we show two examples of (ϕ_1, ϕ_2) with different noise strengths.

In the numerical simulation of Sect. 3.3, we also impose a tiny delay τ for the induced beating, which means that the induced beating of cell j happens at time $t+\tau$ when its neighborhood cell i beats at time t (i.e., if $\phi_i(t-) = 2\pi$ and $\phi_j(t) > B_j$, then $\phi(t) = 0$ and $\phi_j(t+\tau) = 0$).

3.2.4 The CV of Synchronization

For sufficiently small noise, we introduce the synchronization phase function ϕ_i^s for two coupled cardiomyocytes, $i = 1, 2$,

$$\phi_i^s = \mu^s t + \psi_i^s, \tag{3.34}$$

where μ^s is the synchronization frequency for the two cardiomyocytes, and $\{\psi_i^s\}_{i=1,2}$ are two constants. When the noise strengths $\{\sigma_i\}_{i=1,2}$ are small, one can take the synchronization phase function $\{\phi_i^s\}_{i=1,2}$ as the linear approximation of the expected value of $\{\phi_i\}_{i=1,2}$ (see Fig. 3.8).

Without loss of generality, we assume $\psi_1^s \geq \psi_2^s$. Then, the expected value of synchronization period is given by $T^s = (2\pi - \psi_1^s)/\mu^s$.

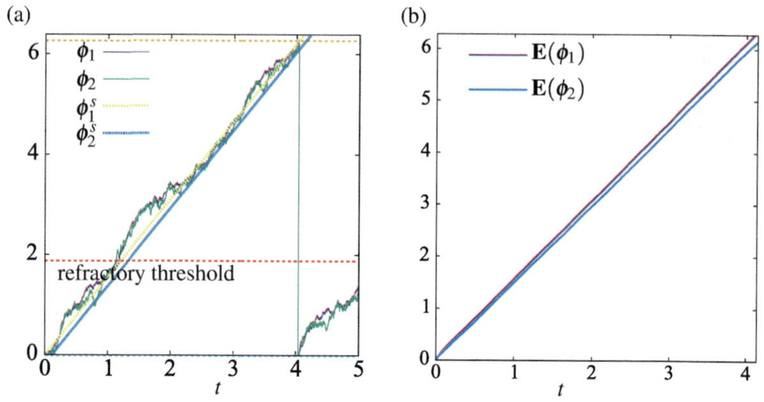

Fig. 3.8 (**a**) A trajectory of $(\phi_1(t), \phi_2(t))$ with parameters $(\mu_1, \mu_2, \sigma_1, \sigma_2) = (1, 2, 0.3, 0.5)$, $A_{1,2} = A_{2,1} = 2$, and $B_1 = B_2 = 0.6\pi$. The synchronization solution (ϕ_1^s, ϕ_2^s) is a linear approximation of $(\mathbf{E}(\phi_1(t)), \mathbf{E}(\phi_2(t)))$. (**b**) $(\mathbf{E}(\phi_1(t)), \mathbf{E}(\phi_2(t)))$

For simplicity of analysis, the effect of reflective boundary at $\phi_i(t) = 0$ is approximated by adding a small positive constant ψ_i^{app} to the initial state:

$$d\phi_i^{\mathrm{app}} = \left[\mu_i + A_{i,j} f\left(\phi_j^{\mathrm{app}} - \phi_i^{\mathrm{app}}\right)\right]dt + \sigma_i dW_i(t), \tag{3.35a}$$

$$\phi_i^{\mathrm{app}}(0) = \psi_i^{\mathrm{app}}. \tag{3.35b}$$

The phase function ϕ_i^{app} is regarded as the approximation of ϕ_i.

Instead of considering the CV of the synchronization period, we investigate the variance of the difference between the phase $\phi_i^{\mathrm{app}}(T^s)$ and the synchronization phase $\phi_i^s(T^s)$ (cf. [29]):

$$\mathbf{CV}_i \approx \sqrt{\mathbf{Var}[\phi_i^{\mathrm{app}}(T^s) - \phi_i^s(T^s)]}.$$

Putting $\xi_i = \phi_i^{\mathrm{app}} - \phi_i^s$,

$$\begin{aligned} f(\phi_j^{\mathrm{app}} - \phi_i^{\mathrm{app}}) &= f(\xi_j - \xi_i + \phi_j^s - \phi_i^s) \\ &= f(\phi_j^s - \phi_i^s) + f'(\phi_j^s - \phi_i^s)(\xi_j - \xi_i) + O(|\xi_j - \xi_i|^2). \end{aligned}$$

Since ξ_i is a small variable when the noise strength is small (see Fig. 3.8a), ignoring the tiny quadratic term $O(|\xi_j - \xi_i|^2)$, and noting that $\phi_j^{\mathrm{app}} - \phi_i^{\mathrm{app}} = \psi_j^s - \psi_i^s$, the above equation becomes

$$f(\phi_j^{\mathrm{app}} - \phi_i^{\mathrm{app}}) = f(\psi_j^s - \psi_i^s) + f'(\psi_j^s - \psi_i^s)(\xi_j - \xi_i).$$

The equations of ξ_i are stated as follows: for $0 < t < T^s$ and $i, j = 1, 2, i \neq j$,

$$d\xi_i = \nu_i dt + A_{i,j} f'(\psi_j^s - \psi_i^s)(\xi_j - \xi_i)dt + \sigma_i dW_i, \tag{3.36a}$$

$$\xi_i(0) = \xi_i^0, \tag{3.36b}$$

where $\nu_i = \mu_i - \mu^s + A_{i,j} f(\psi_j^s - \psi_i^s)$ and $\xi_i^0 = \psi_i^{\text{app}} - \psi_i^s$.

In view of

$$\xi_i(T^s) - \xi_i(0) = \phi_i^{\text{app}}(T^s) - \phi_i^{\text{app}}(0) - (\phi_i^s(T^s) - \phi_i^s(0)),$$

$\mathbf{CV}_i \approx \sqrt{\mathbf{Var}[\phi_i^{\text{app}}(T^s) - \phi_i^s(T^s)]}$ is equivalent to

$$\mathbf{CV}_i \approx \sqrt{\mathbf{Var}[\xi_i(T^s) - \xi_i^0]} = \sqrt{\mathbf{Var}[\xi_i(T^s)]}.$$

Setting the notations

$$\boldsymbol{\xi} = \begin{bmatrix} \xi_1 \\ \xi_2 \end{bmatrix}, \quad \boldsymbol{\nu} = \begin{bmatrix} \nu_1 \\ \nu_2 \end{bmatrix}, \quad \boldsymbol{\xi}^0 = \begin{bmatrix} \xi_1^0 \\ \xi_2^0 \end{bmatrix}, \quad \boldsymbol{W} = \begin{bmatrix} W_1 \\ W_2 \end{bmatrix},$$

$$\boldsymbol{B} = \begin{bmatrix} b_{12} & -b_{12} \\ -b_{21} & b_{21} \end{bmatrix}, \quad \boldsymbol{\sigma} = \begin{bmatrix} \sigma_1 & 0 \\ 0 & \sigma_2 \end{bmatrix},$$

where $b_{ij} = A_{i,j} f'(\psi_j^s - \psi_i^s)$, we rewrite (3.36) into

$$d\boldsymbol{\xi} = \boldsymbol{\nu} dt - \boldsymbol{B}\boldsymbol{\xi} dt + \boldsymbol{\sigma} d\boldsymbol{W}(t).$$

Next, we compute the solution $\boldsymbol{\xi}$. Multiplying the above equation by $e^{-\boldsymbol{B}t}$, we get

$$d\left(e^{\boldsymbol{B}t}\boldsymbol{\xi}\right) = e^{\boldsymbol{B}t}[\boldsymbol{\nu} dt + \boldsymbol{\sigma} d\boldsymbol{W}],$$

which yields

$$\boldsymbol{\xi}(t) = e^{-\boldsymbol{B}t}\boldsymbol{\xi}^0 + \int_0^t e^{-\boldsymbol{B}(t-s)}\boldsymbol{\nu}\, ds + \int_0^t e^{-\boldsymbol{B}(t-s)}\boldsymbol{\sigma}\, d\boldsymbol{W}(s). \tag{3.37}$$

By Itô's integral (3.26a), we calculate

$$\mathbf{E}(\boldsymbol{\xi}(t)) = [\mathbf{E}(\xi_1(t)), \mathbf{E}(\xi_2(t))]^\top = e^{-\boldsymbol{B}t}\boldsymbol{\xi}^0 + \int_0^t e^{-\boldsymbol{B}(t-s)}\boldsymbol{\nu}\, ds,$$

$$\mathbf{E}((\xi_i(t))^2) = [\mathbf{E}(\xi_i(t))]^2 + \mathbf{E}\left[\int_0^t \left[e^{-(t-s)\boldsymbol{B}}\boldsymbol{\sigma} d\boldsymbol{W}(s)\right]_i\right]^2,$$

where $[\boldsymbol{u}]_i$ denotes the ith component of vector $\boldsymbol{u}$. Together with $\mathbf{Var}[\xi_i(t)] = \mathbf{E}[(\xi_i(t))^2] - (\mathbf{E}[\xi_i(t)])^2$, we have

$$\mathbf{Var}[\xi_i(t)] = \mathbf{E}\left[\int_0^t \left[e^{-(t-s)\boldsymbol{B}}\boldsymbol{\sigma} d\boldsymbol{W}\right]_i\right]^2.$$

Noting that $\boldsymbol{B}$ has two eigenvalues

$$\lambda_1 = 0, \quad \lambda_2 = b := b_{12} + b_{21},$$

with the corresponding eigenvectors

$$\boldsymbol{u}_1 = [1, 1]^\top, \quad \boldsymbol{u}_2 = [b_{12}, -b_{21}]^\top,$$

we introduce the decomposition

$$\boldsymbol{\sigma} d\boldsymbol{W} = b^{-1}(b_{21}\sigma_1 dW_1 + b_{12}\sigma_2 dW_2)\boldsymbol{u}_1 + b^{-1}(\sigma_1 dW_1 - \sigma_2 dW_2)\boldsymbol{u}_2.$$

Putting together with

$$e^{-t\boldsymbol{B}}\boldsymbol{u}_1 = e^{-t0}\boldsymbol{u}_1 = \boldsymbol{u}_1, \quad e^{-t\boldsymbol{B}}\boldsymbol{u}_2 = e^{-tb}\boldsymbol{u}_2,$$

one can compute $[e^{-(t-s)\boldsymbol{B}}\boldsymbol{\sigma} d\boldsymbol{W}]_i$. Then, applying Itô's integral (3.26b), it is not difficult to validate that

$$\begin{aligned}\mathbf{CV}_i^2 &\approx \mathbf{Var}[\xi_i(T^s)] \\ &= \frac{1}{b^2}\left[\left(b_{12}^2\sigma_2^2 + b_{21}^2\sigma_1^2\right)T^s + \frac{1-e^{-2bT^s}}{2b} b_{ij}^2\left(\sigma_1^2 + \sigma_2^2\right)\right. \\ &\quad \left. + \frac{1-e^{-bT^s}}{b} 2b_{ij}\left(b_{ji}\sigma_i^2 - b_{ij}\sigma_j^2\right)\right].\end{aligned} \tag{3.38}$$

As stated above (cf. [29]), one can approximate the CV of synchronization period by $\mathbf{Var}[\xi_i(T^s)]$.

To reduce the difference between $\mathbf{CV}_1$ and $\mathbf{CV}_2$, we take $b_{21}\sigma_1^2 = b_{12}\sigma_2^2 = D$ such that

$$\mathbf{CV}_i^2 \approx \frac{1}{b^2}\left[2DT^s + \frac{1-e^{-2bT^s}}{2b}\frac{D^2}{\sigma_j^4}\left(\sigma_1^2 + \sigma_2^2\right)\right].$$

Furthermore, for $\sigma_1 = \sigma_2 = \sigma$ and $b_{21} = b_{12} = b/2$, we have

$$\mathbf{CV}_1^2 = \mathbf{CV}_2^2 \approx \frac{\sigma^2}{2}\left[T^s + \frac{1 - e^{-2bT^s}}{2b}\right],$$

which is regarded as the CV of the synchronization period.

In numerical simulation, we set $f(x) = \sin(x)$. Noting that $\psi_i^s \approx 0$ for small noise, we have $f'(\psi_j^s - \psi_i^s) \approx \cos(0) = 1$ such that $b_{ij} \approx -A_{i,j}$. Therefore, choosing $A_{i,j}\sigma_j^2 = A_{j,i}\sigma_i^2$, one can reduce the difference between $\mathbf{CV}_1$ and $\mathbf{CV}_2$, which yields a more stable synchronization solution in simulation. This has been applied to our numerical simulation in Sect. 3.3.

3.2.5 The Phase Model for N-cardiomyocytes Network

For N-cardiomyocytes network, we need to consider the neighborhood cells of each cell. For simplicity, we consider the full connected network, where the cells are connected with each other. The equations of $\{\phi_i\}_{i=1}^N$ are stated as follows:

$$d\phi_i(t) = \mu_i dt + \sum_{j=1,\ j\neq i}^{N} A_{i,j} f(\phi_j - \phi_i)dt + \sigma_i dW_i(t) + dL_i(t), \tag{3.39a}$$

$$\phi_i(0) = 0. \tag{3.39b}$$

In (3.39a), the reaction term $A_{i,j} f(\phi_j - \phi_i)$ between cells i and j is imposed for $j = 1, 2, \ldots, N,\ j \neq i$, which corresponds to the full connection of the cell network. The process $L_i(t)$ is added to (3.39a) to guarantee that $\phi_i(t) \geq 0$ always holds even when the noise dW_i is negative and $\phi_i(t) = 0$, which represents the irreversibility of beating (see (i) and (ii) of Sect. 3.2.2).

Moreover, we also incorporate the conception of the induced beating and refractory to the mathematical model: for $i, j = 1, \ldots, N$ and $i \neq j$,

(R1) when the cardiac muscle cell i beats (i.e., $\phi_i(t-) = 2\pi$) and the cell j is out of refractory (i.e., $\phi_j(t-) > B_j$), then both the cells beat and the phase of cells i and j returns to 0 (i.e., $\phi_i(t) = 0$ and $\phi_j(t) = 0$).

(R2) when the cardiac muscle cell i beats (i.e., $\phi_i(t-) = 2\pi$) and the cell j is in refractory (i.e., $\phi_j(t-) \leq B_j$), then only the cell i is beating and the phase of cell i returns to 0 (i.e., $\phi_i(t) = 0$ and $\phi_j(t) = \phi_j(t-)$).

In Fig. 3.9, we show two examples of $(\phi_1, \phi_2, \phi_3, \phi_4)$ with different noise strengths.

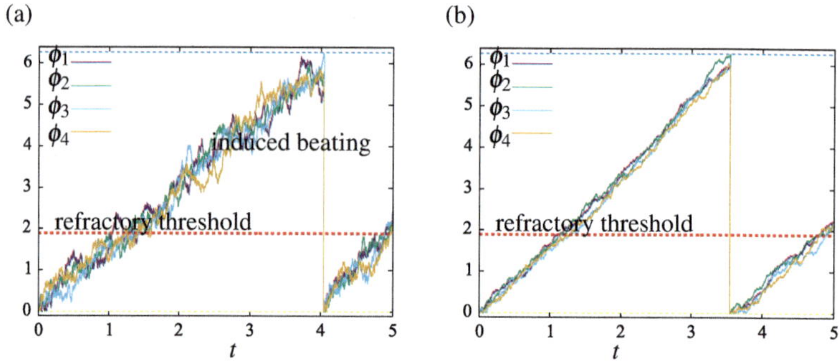

Fig. 3.9 The trajectories of $(\phi_1(t), \phi_2(t), \phi_3(t), \phi_4(t))$ with parameters $(\mu_1, \mu_2, \mu_3, \mu_4) = (2, 2, 1, 1)$ and $A_{i,j} = 2$, $B_i = 0.6\pi$ for $i, j = 1, \dots, 4$, $i \neq j$. (**a**) The strong noise strength $\sigma_i = 1$. (**b**) The weak noise strength $\sigma_i = 0.3$

As with the case of two coupled cells, we investigate the CV of the synchronization period using the approximated calculation. For sufficiently small noise, let ϕ_i^{app} denote the approximation of the phase function ϕ_i, satisfying: $1 \leq i \leq N$,

$$d\phi_i^{\text{app}} = \left[\mu_i + \sum_{j=1, j\neq i}^{N} A_{i,j} f\left(\phi_j^{\text{app}} - \phi_i^{\text{app}}\right) \right] dt + \sigma_i dW_i(t), \tag{3.40a}$$

$$\phi_i^{\text{app}}(0) = \psi_i^{\text{app}}. \tag{3.40b}$$

If the noise is sufficiently small, then we can approximate the synchronization by linear equations:

$$\phi_i^s = \mu^s t + \psi_i^s, \quad 1 \leq i \leq N. \tag{3.41}$$

Without loss of generality, we assume $\psi_1^s = \max_{1\leq i \leq N} \psi_i^s$. Then, the synchronization period is given by $T^s = (2\pi - \psi_1^s)/\mu^s$. Introducing the difference $\xi_i = \phi_i^{\text{app}} - \phi_i^s$, we introduce the approximated CV of the synchronization (cf. [29]): $\mathbf{CV}_i \approx \sqrt{\mathbf{Var}[\xi_i(T^s) - \xi_i(0)]} = \sqrt{\mathbf{Var}[\xi_i(T^s)]}$. In the following, we calculate $\mathbf{Var}[\xi_i(T^s)]$.

First, we see that $\{\xi_i\}_{i=1}^N$ satisfy the following equations: $i = 1, \dots, N$,

$$d\xi_i = \left[\nu_i + \sum_{j=1, j\neq i}^{N} A_{i,j} f'(\psi_j^s - \psi_i^s)(\xi_j - \xi_i) \right] dt + \sigma_i dW_i, \tag{3.42a}$$

$$\xi_i(0) = \xi_0 \tag{3.42b}$$

where $\nu_i = \mu_i - \mu^s + \sum_{j=1,j\neq i}^{N} A_{i,j} f(\psi_j^s - \psi_i^s)$ and $\xi_i^0 = \psi_i^{\text{app}} - \psi_i^s$. Setting the notations

$$\boldsymbol{\xi} = [\xi_1, \dots, \xi_N]^\top, \quad \boldsymbol{W} = [W_1, \dots, W_N]^\top, \tag{3.43a}$$

$$\boldsymbol{\nu} = [\nu_1, \dots, \nu_N]^\top, \quad \boldsymbol{\xi}^0 = [\xi_1^0, \dots, \xi_N^0]^\top, \tag{3.43b}$$

$$\boldsymbol{\sigma} = \mathrm{diag}(\sigma_1, \dots, \sigma_N), \quad \boldsymbol{B} = [b_{ij}]_{1\leq i,j\leq N}, \tag{3.43c}$$

where $b_{ii} = \sum_{j=1,j\neq i}^{N} A_{i,j} f'(\psi_j^s - \psi_i^s)$ and $b_{ij} = -A_{i,j} f'(\psi_j^s - \psi_i^s)$, we rewrite (3.42) into

$$d\boldsymbol{\xi} = \boldsymbol{\nu} dt - \boldsymbol{B}\boldsymbol{\xi} dt + \boldsymbol{\sigma} d\boldsymbol{W},$$
$$\boldsymbol{\xi}(0) = \boldsymbol{\xi}_0.$$

The solution $\boldsymbol{\xi}$ is given by

$$\boldsymbol{\xi}(t) = e^{-\boldsymbol{B}t}\boldsymbol{\xi}^0 + \int_0^t e^{-\boldsymbol{B}(t-s)}\boldsymbol{\nu}\, ds + \int_0^t e^{-\boldsymbol{B}(t-s)}\boldsymbol{\sigma}\, d\boldsymbol{W}(s).$$

Putting $|\boldsymbol{\xi}|^2 = \sum_{i=1}^{N} |\xi_i|^2$, we see that

$$|\boldsymbol{\xi}(t)|^2 = |\boldsymbol{\xi}^0|^2 + 2\boldsymbol{\xi}^0 \cdot \int_0^t e^{-(t-s)\boldsymbol{B}}\boldsymbol{\sigma} d\boldsymbol{W} + \left|\int_0^t e^{-(t-s)\boldsymbol{B}}\boldsymbol{\sigma} d\boldsymbol{W}\right|^2.$$

To obtain $\mathbf{Var}(\xi(T^s))$, we introduce the properties of Itô's formula for the N-dimensional version [6, 35]: for any $\boldsymbol{G} = [G_{ij}]_{i,j}$ satisfying $\int_0^t |\boldsymbol{G}|^2\, ds < \infty$ ($|\boldsymbol{G}|^2 = \sum_{i,j=1}^{N} |G_{ij}|^2$), the following equalities hold:

$$\mathbf{E}\left(\int_0^t \boldsymbol{G}(s)\, d\boldsymbol{W}(s)\right) = 0, \tag{3.44a}$$

$$\mathbf{E}\left[\left|\int_0^t \boldsymbol{G}(s)\, d\boldsymbol{W}(s)\right|^2\right] = \mathbf{E}\left(\int_0^t |\boldsymbol{G}|^2\, ds\right). \tag{3.44b}$$

Applying (3.44),

$$\mathbf{E}[\boldsymbol{\xi}(t)] = e^{-\boldsymbol{B}t}\boldsymbol{\xi}^0 + \int_0^t e^{-\boldsymbol{B}(t-s)}\boldsymbol{\nu}\, ds, \tag{3.45a}$$

$$\mathbf{E}[|\boldsymbol{\xi}(t)|^2] = (\mathbf{E}[\boldsymbol{\xi}(t)])^2 + \mathbf{E}\left[\int_0^t \left|e^{-(t-s)\boldsymbol{B}}\boldsymbol{\sigma}\right|^2 ds\right]. \tag{3.45b}$$

For the case $\sigma_i = \sigma$, $A_{i,j} = A_{j,i}$, and $\psi_i^s = \psi^s$, we have

$$\begin{aligned}\left|e^{-(t-s)\boldsymbol{B}}\boldsymbol{\sigma}\right|^2 &= \sigma^2 \sum_{i,j=1}^{N} \left|\left(e^{-(t-s)\boldsymbol{B}}\right)_{ij}\right|^2 \\ &= \mathrm{tr}\left(\left(e^{-(t-s)\boldsymbol{B}}\right)^\top e^{-(t-s)\boldsymbol{B}}\right) = \mathrm{tr}\left(e^{-2(t-s)\boldsymbol{B}}\right) \quad (\text{by } \boldsymbol{B}^\top = \boldsymbol{B}) \\ &= \sum_{i=1}^{N} e^{-2(t-s)\lambda_i},\end{aligned} \tag{3.46}$$

where $\mathrm{tr}(\boldsymbol{B})$ and $\{\lambda_i\}_{i=1}^N$ denote the trace and the eigenvalues of $\boldsymbol{B}$, respectively.

Following from $A_{i,j} = A_{j,i}$ and $f'(x) = f'(-x)$ (by $f(x) = -f(-x)$), the symmetry $\boldsymbol{B}^\top = \boldsymbol{B}$ holds. Without loss of generality, let $\lambda_N \geq \lambda_{N-1} \geq \cdots \geq \lambda_2 > \lambda_1 = 0$ (0 is an eigenvalue of $\boldsymbol{B}$ because of $\sum_{j=1}^N b_{ij} = 0$). We obtain from (3.45), (3.46), and $\sum_{i=1}^N \mathbf{Var}[\xi_i(t)] = \mathbf{E}[|\boldsymbol{\xi}(t)|^2] - |\mathbf{E}[\boldsymbol{\xi}(t)]|^2$ that

$$\begin{aligned}\sum_{i=1}^{N} \mathbf{Var}[\xi_i(T^s)] &= \sigma^2 \int_0^{T^s} \sum_{i=1}^{N} e^{-2(T^s-s)\lambda_i}\, ds. \\ &= \sigma^2 T^s + \sigma^2 \sum_{i=2}^{N} \frac{1-e^{-2T^s\lambda_i}}{2\lambda_i}.\end{aligned} \tag{3.47}$$

For the case of the identical parameter $(\mu_i, \sigma_i, A_{i,j}) = (\mu, \sigma, A)$, ξ_i has identical distribution for each i, as well as $\sqrt{\mathbf{Var}[\xi_i(T^s)]}$. As a result, we have, for all $i = 1, 2, \ldots, N$,

$$\begin{aligned}\mathbf{CV}_i \approx \sqrt{\mathbf{Var}[\xi_i(T^s)]} &= \sqrt{\sum_{i=1}^{N} \mathbf{Var}[\xi_i(T^s)/N]} \\ &= \frac{\sigma}{\sqrt{N}} \sqrt{T^s + \sum_{i=2}^{N} \frac{1-e^{-2T^s\lambda_i}}{2\lambda_i}}.\end{aligned}$$

Noting that there exists a constant c_∞ such that

$$\frac{1}{N} \sum_{i=2}^{N} \left(\frac{1-e^{-2T^s\lambda_i}}{2\lambda_i}\right) \to c_\infty \quad \text{as } N \to \infty, \tag{3.48}$$

we have $\mathbf{CV}_i \to \sigma c_\infty^{1/2}$ as $N \to \infty$.

In Sect. 3.4, the numerical simulations are carried out for several N-cell models with various network, where we observe that the CV decreases by $O(1/\sqrt{N})$ when N is small and converges to some constant when $N \to \infty$. The simulations correspond to the theoretical result (3.48). Moreover, the comparison between the numerical simulation and biological experimental result (cf. [18, Figure 3]) indicates the well consistency between the model and biological experiment.

3.3 Experimental Approach

3.3.1 On-Chip Cellomics Technology: Reconstructive Understanding of the Community Effect in Cardiomyocytes

We have developed a constructive experimental approach for understanding epigenetic information. As shown in Fig. 3.10, the strategy behind our on-chip microfabrication methods, which we call "on-chip cellomics technologies" [58], is constructed through three steps. First, we purify target cells from tissue individually in a nondestructive manner using several technologies, such as digestible DNA-aptamer labeling and cell collection, ultrahigh-speed camera-based real-time imaging cell sorting, or noninvasive handling of cells using an acoustic radiation force [1, 12, 57, 59–61]. We then cultivate and observe the cells under fully controlled conditions (e.g., cell population, network patterns, or nutrient conditions) using an on-chip single-cell cultivation chip [14, 15, 21, 34, 50–54] or an on-chip agarose microchamber system exploiting photothermal etching technology, which can control the microstructure of microchambers even during cell cultivation [11, 24, 24–28, 36, 43–49]. Finally, we undertake single-cell-based genome/proteome analysis through a set of nanoprobes and adaptive electron microscopy, single-cell-based DNA/RNA release technology, or a 3-min ultrahigh-speed polymerase chain reaction (PCR) measurement technology [62].

The advantage of the experimental on-chip cellomics approach is that, as it is a reconstructive approach of the simplified artificial minimum cell network model on a chip, it removes the complexity of the underlying physicochemical reactions that are not always completely understood and for which most of the necessary variables cannot be measured. Moreover, this approach shifts the view of cell regulatory processes from basic chemical grounds to a paradigm of the cell as an information processing unit working as an intelligent machine capable of adapting to changing environmental and internal conditions. This is an alternative representation of the cell and can provide new insights into cellular processes. Thus, models derived from such a viewpoint can directly help in more conventional biochemical and molecular biological analyses that assist in our understanding of control in cells.

From the geometric perspective, two more detailed viewpoints of analysis should also be taken: one is on the population/community size dependence and the other

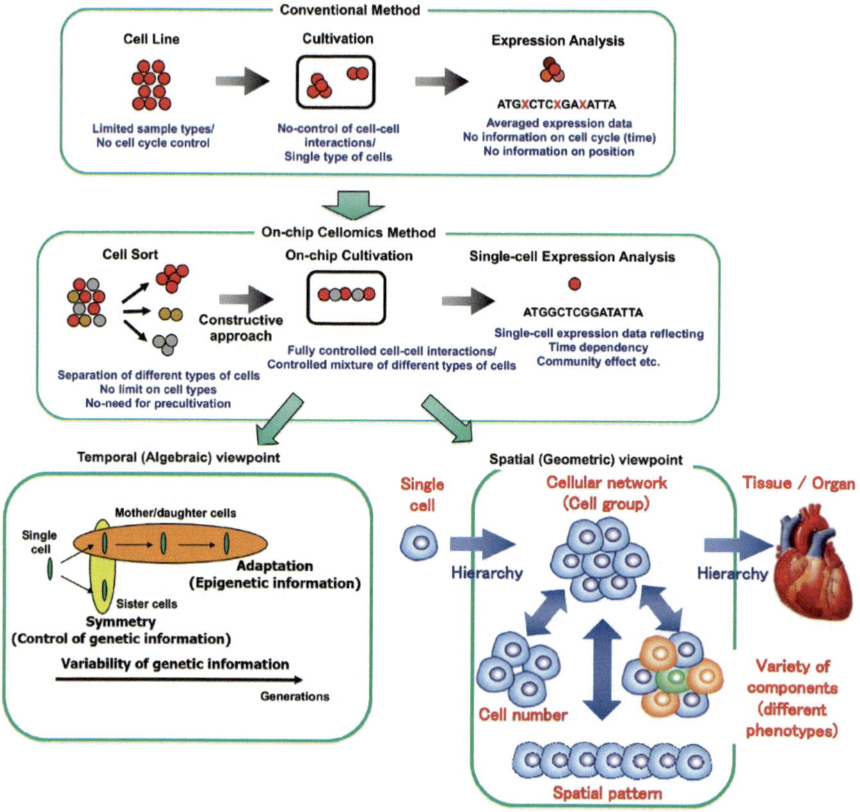

Fig. 3.10 On-chip cellomics analysis. The aim of single-cell-based analysis of multicellular systems: temporal (algebraic) aspect and spatial (geometric) aspect

is on the spatial (network) pattern dependence of groups of cells. In conventional cell-based studies, cell lines are usually used for acquiring the same type of cells and are then cultivated in a cultivation dish without any control of their population or any formation of a community with other cell types. Finally, they are analyzed as a group regardless of any differences in their cell cycle regardless of their possible differences. In contrast, on-chip cellomics technology involves a new strategy with three steps: First, the cells are taken from a community using a nondestructive cell sorting procedure. Then, the cells are cultivated in a microchamber, in which cell network formation and medium environment are controlled. Finally, the genome/proteome measurement in each cell is measured (Fig. 3.10).

3.3.2 Photothermal Etching on Agarose Layer for Cell Network Formation Control

Flexible change of microstructures of cell-to-cell interactions or cell network shapes on a chip during cultivation is necessary for the "temporal" and "geometric" reconstructive approach of cell network studies. To accomplish this requirement, we have developed a photothermal etching method [11, 24, 26, 36, 46] with an agarose microchamber cell cultivation system (Fig. 3.11). This involves the area-specific melting of a portion of agarose of a whole light pathway by spot heating using a focused infrared laser beam of 1480 nm, which absorbs water, and of a portion of agarose close to a thin layer made of a light-absorbing material, such as chromium, with a laser beam of 1064 nm, which is permeable to water. When we combine infrared lasers with these two different wavelengths, we can fabricate microchambers and microtunnels flexibly for the noncontact three-dimensional photothermal etching of agarose. In other words, as the 1480-nm infrared beam is associated with the absorption of water and agarose gel, the agarose gel in the 1480-nm infrared light pathway was heated and completely melted. Moreover, as the 1064-nm infrared beam was not associated with this absorbance, the agarose melted just near the thin chromium layer, which absorbed the beam.

For phase-contrast microscopy and this μm-scale photothermal etching, light of three different wavelengths (visible light for observation and 1480-nm/1064-nm infrared lasers for spot heating to construct microchambers/microtunnels, respectively) was used simultaneously to observe the positions of the agarose chip surface and to melt a portion of the agarose in the area being heated. As described above, the advantage of this method is that we can apply this stepwise network formation (addition) approach even during cultivation, so we can change the network size and pattern of cardiomyocyte cells during cultivation by adding microchannels between two adjacent microchambers in a step-by-step fashion; moreover, this approach is also applicable for neuronal networks [27, 28, 43, 45–49].

3.3.3 Community Effect of Cells for Their Synchronization (1): Two-Cell Model

As described in the previous subsection, the ability of photothermal etching of agarose microstructures to control the cell arrangement is beneficial for cardiomyocyte network studies. In this subsection, we introduce the application of this technology to reveal the involvement of the community effect in cardiomyocyte beating synchronization[17, 17–20, 24–28].

First, we investigated the roles of the beat rates (interbeat intervals) and beat rate fluctuation of isolated single cardiomyocytes in the reestablishment of synchronous beating by analyzing the changes of beating rates and their fluctuations before and after the synchronization of two cardiomyocytes through narrow channels with initially different rhythms (e.g., Fig. 3.12A and B)[27, 28]. The results showed three

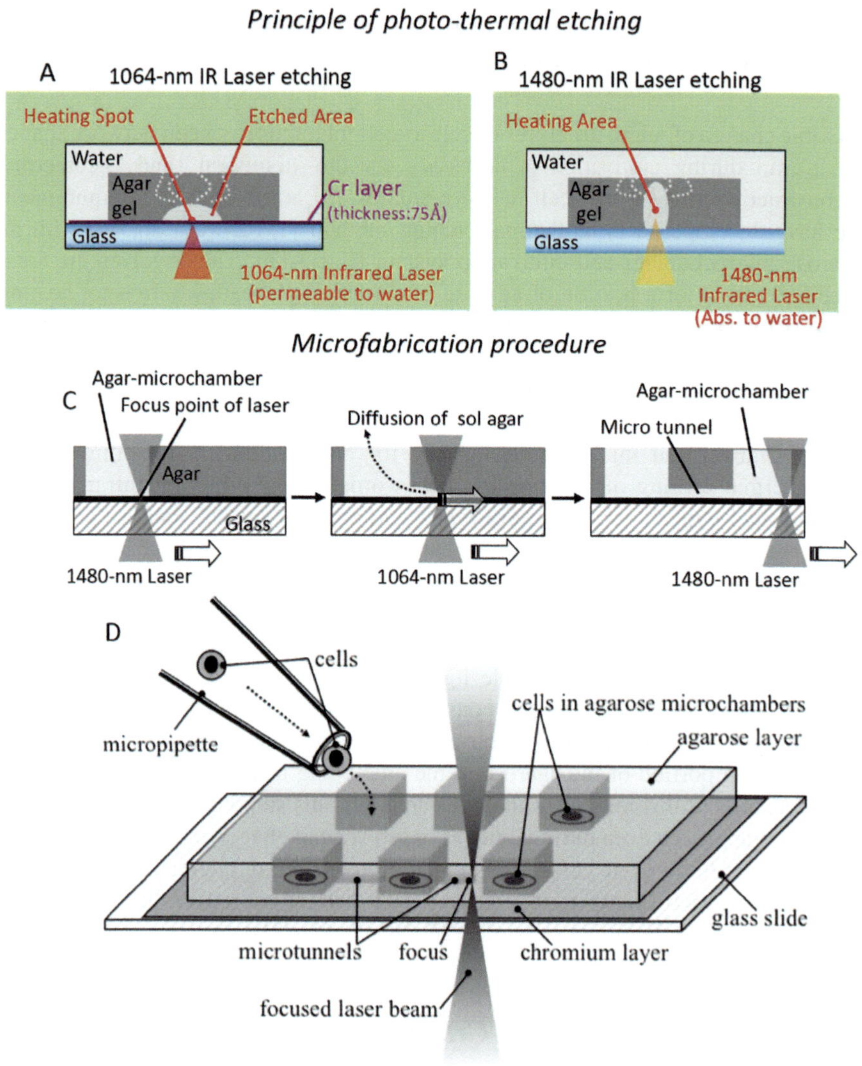

Fig. 3.11 Photothermal etching method. Using focused infrared (IR) lasers of two different wavelengths, the thin layer of low-melting point agarose on the chip was selectively melted in different manners. (**a**) As the 1064-nm IR laser is not associated with the absorption of water, only a portion of the agarose near the thin absorption layer is heated and melted, changing its state from a gel to a sol. In contrast, (**b**) as the 1480-nm IR laser is associated with the absorption of water, all of the agarose in the light pathway is heated and melted. (**c**) The agarose changed to a sol state is dispersed into the agarose gel, holes or tunnels are formed in the agarose layer, and (**d**) cells are inserted in agarose microchambers with a micropipette

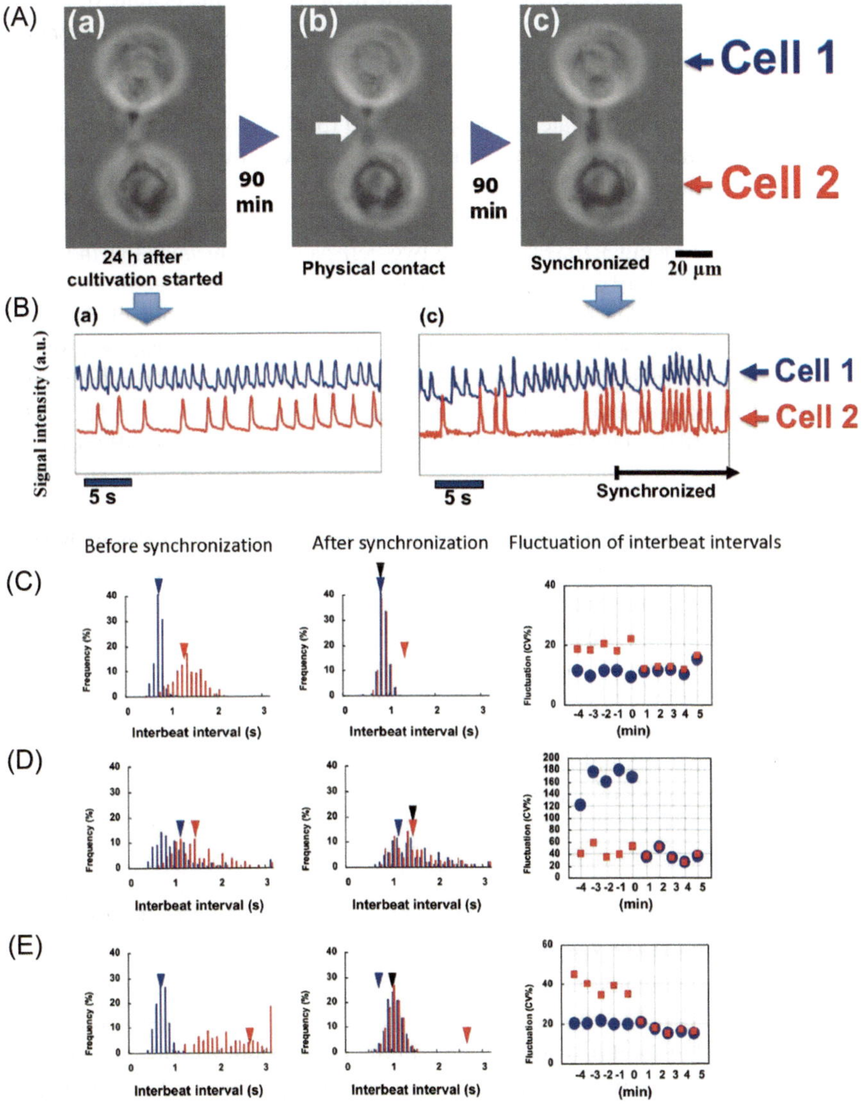

Fig. 3.12 Synchronization of two cardiomyocytes. (**A**) Micrographs of two cardiomyocytes under isolated conditions (a), just after they were connected together (b), and just after synchronization started (c). (**B**) Beating waveforms at (a) and (c) in (**A**). (**C**)–(**E**) (left graph and center graph) Beating frequency spectrum before and after synchronization, respectively; distribution of interbeat intervals of two cardiomyocytes, and the change of the mean value of beating rhythm fluctuation at intervals of 1 min measured for 5 min before and after synchronization. Blue and red triangles show the mean values before synchronization, and black triangles show the mean value for the two cells after synchronization. (right graph) The change of the mean value of beating rhythm fluctuation [CV%: coefficient of variation (100 × standard deviation/mean beat rate)] at intervals of 1 min measured for 5 min before and after synchronization. Blue circles and red squares show the corresponding mean values of beating rhythm fluctuation for 1 min. Three types of synchronization tendencies were described: (**C**) synchronization to a faster beating cell, (**D**) synchronization to a slower beating cell, and (**E**) synchronization with a new beating frequency

types of synchronization of two cardiomyocyte networks: (1) the beating of the two cardiomyocytes synchronized at the faster of the two initial rates, but there was beating fluctuation at the lower of the two initial rates (Fig. 3.12C), (2) the beating of the two cells synchronized at the lower of the two initial rates but fluctuated at the lower of the two initial rates (Fig. 3.12D), and (3) the synchronization occurred at neither of the initial rates of single cardiomyocytes, with fluctuation of smaller of the initial fluctuations (Fig. 3.12E).

The interbeat intervals of 14 two-cell pairs before and after synchronization are listed in Table a of Fig 3.13. Five of the two-cell pairs synchronized at the initial rate of the faster cell, two of the pairs synchronized at the initial rate of the slower cell, and the other seven pairs synchronized at a rate other than one of the initial rates. In Table b of Fig. 3.13, the fluctuation data for the 14 cell pairs whose rate data are listed in Table A are grouped according to the change of the fluctuation before and after synchronization. Thirteen pairs synchronized with a fluctuation equal to or less than the initial fluctuation of the slower member of the pair, and one pair synchronized with a fluctuation larger than that of either of the two initial fluctuations.

These results suggest that the fluctuation of reestablished synchronous beating by isolated cardiomyocytes is influenced more strongly by the fluctuation of the initial fluctuation of the beat rates of the isolated cardiomyocytes than the rate of the reestablished synchronous beating is influenced by the initial beat rates of the isolated cardiomyocytes. It is therefore possible that a cardiomyocyte whose beat rate fluctuates less than that of another cardiomyocyte entrains the beating rhythm of that cardiomyocyte, but we observed one pair of cells in which this did not happen. This indicates that the influence of a single cell is still not sufficiently strong to account for the process of entrainment in heart tissue.

3.3.4 Community Effect of Cells for Their Synchronization (2): Cell Number Dependence

Figure 3.14 also describes the community size effect of a cardiomyocyte network on its beating stability. In this work, we explore the relationship between entrainment and community size by examining the synchronization process of a cardiomyocyte network formed by the interaction of single cardiomyocytes cultured in a 3×3 grid of agarose microchambers with connecting microchannels[28]. After nine isolated cells had been cultured in the nine-chamber agarose microcultivation chip for 24 h, we started to measure the synchronization process continuously and found that, when an isolated single cell came into contact with another cell and formed a two-cell network (Fig. 3.14 top and middle), these two cells synchronized at the initial rate of the first cell and the fluctuation decreased from the initial fluctuation of the first cell. When all nine cells came into contact and formed a nine-cell network (Fig. 3.14 bottom), it synchronized at a rate equal to the initial rate of the first cell, with a decrease of fluctuation.

A. Three types of synchronization from the viewpoint of frequency (s) (n = 14)

Synchronize to faster beating cell			Synchronize to slower beating cell			Forming new beating interval		
Before		After	Before		After	Before		After
Cell 1	Cell 2	Sync	Cell 1	Cell 2	Sync	Cell 1	Cell 2	Sync
0.64	1.23	0.76	1.10	1.40	1.40	0.64	2.7	0.94
0.93	1.01	0.83	0.59	0.63	0.62	0.84	1.77	1.10
0.74	1.13	0.82				0.56	1.21	0.89
0.87	1.43	0.86				0.71	0.92	0.81
0.94	2.18	0.89				0.56	1.21	0.90
						0.53	1.06	0.74
						0.43	3.10	1.00

B. Two types of synchronization from the viewpoint of fluctuation (CV%)

Synchronize to fluctuation (CV%) decrease				Synchronize to fluctuation (CV%) increase			
Before		After		Before		After	
Cell 1	Cell 2	Sync		Cell 1	Cell 2	Sync	
25.1	12.3	12.3	Sync to faster beating cell	88.7	26.7	78.1	Forming new beating interval
20.8	16.4	8.9					
46.5	19.7	15.9					
117	19.7	18.9					
164	14.7	13.2					
149	41.2	41.7	Sync to slower beating cell				
16.4	11.6	10.7					
42.9	20.1	17.3	Forming new beating interval				
19.7	11.7	10.9					
29.0	17.9	18.1					
20.5	17.5	12.9					
29.3	17.9	18.7					
22.8	21.7	11.8					

Fig. 3.13 Tendency of synchronization of two cardiomyocytes. (**a**) Three types of synchronization of two cardiomyocytes from the perspective of beating intervals. (**b**) Two types of synchronization from the perspective of beating stability (fluctuation of beating)

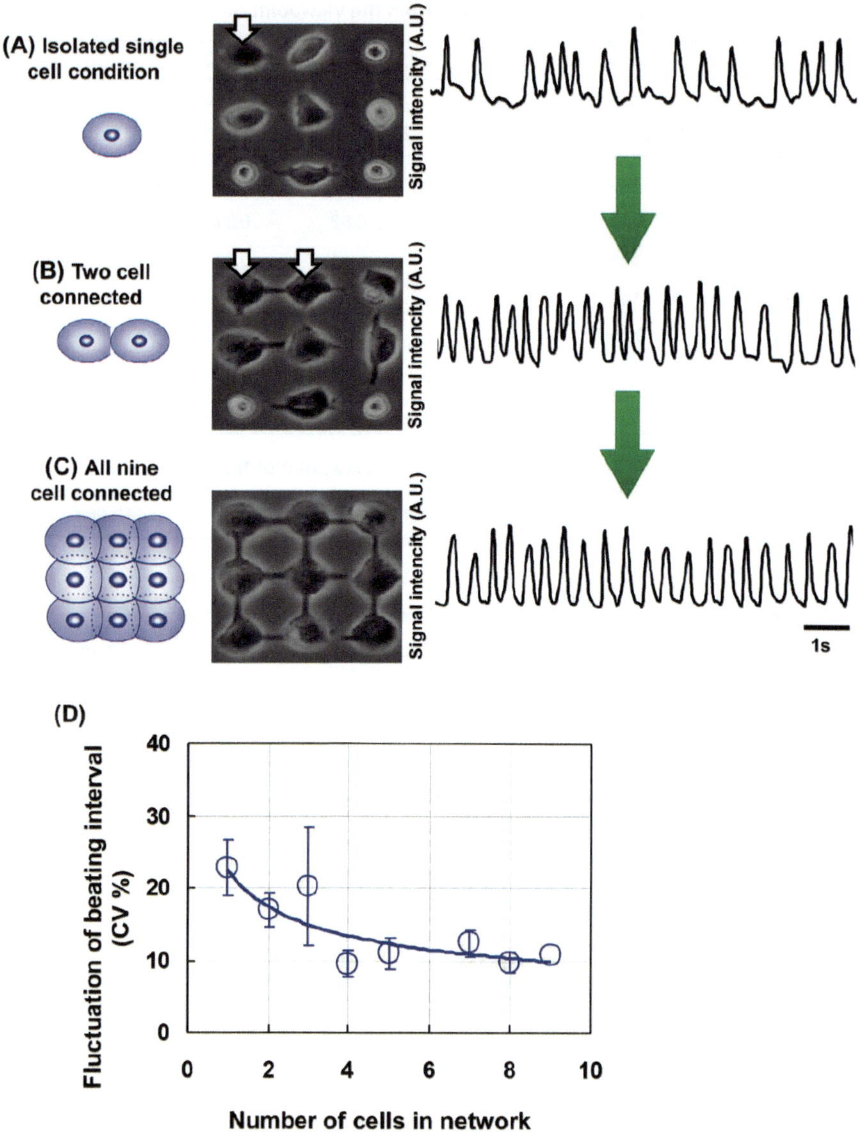

Fig. 3.14 Effect of increase in connected cell number on increase in beating stability. (**a**) Isolated single cell, (**b**) two-cell network, and (**c**) nine-cell network. (**d**) Dependence of beating interval fluctuation on cell number

These results suggest that the beating rhythm of a single cardiomyocyte tends to entrain the rhythm of the cell network, and the strength of this tendency increases with the size of the network. Therefore, it is thought that the fluctuation of the rate at which a network of cardiomyocytes beats decreases as the size of the network increases. The tendency of the synchronization above was simply explained by asserting that the synchronization of two cardiomyocytes was caused by the more unstable cell (the one with the more variable beating intervals) following the more stable cell. Such tendency for reduced fluctuation was more pronounced when the number of cardiomyocytes in the network increased; we call this phenomenon the "community effect" of synchronization.

Using the agarose microchambers, we can examine the dependence on the spatial arrangement of the synchronization stability of cardiomyocyte networks [18]. As shown in Fig. 3.15, we can arrange the cardiomyocytes in three different shapes, a linear shape, a radial shape, and a lattice shape and compare their tendencies for beating stabilization relative to cell numbers. The results indicated that there was no apparent relationship between the number of cells and their shape and that the most important index for the stabilization of cell beating is not the geometry of cells but their number.

3.3.5 Community Effect of Cells on Their Synchronization (3): Mixture of Different Types of Cells

We also examined the contribution of fibroblasts to the synchronization of cardiomyocytes[19]. We connected two cardiomyocytes through a single fibroblast and synchronized them, as shown in Fig. 3.16A and B, and then used this heterogeneous cardiomyocyte–fibroblast coupling to examine the tendency of the stability of interbeat intervals and beating rhythm fluctuation of two cardiomyocytes through a fibroblast before and after their synchronization.

The first type of synchronization involved the tendency for the fluctuation to decrease due to synchronization, which is the same tendency as seen in a network formed by the direct connection of two cardiomyocytes. As shown in Fig. 3.16C and D, in this type, the two cells having different interbeat intervals before synchronization synchronized to achieve an interbeat interval of less than a second after synchronization (e.g., Fig. 3.16B). The fluctuation of the synchronized network became almost equal to or smaller than either of the two initial fluctuations.

In contrast, the second type involved the tendency for the fluctuation to increase due to synchronization, which did not occur in the cardiomyocyte network (Fig. 3.16E). In this case, two cardiomyocytes having two different interbeat intervals before synchronization exhibit a higher mean interbeat interval after synchronization, and the fluctuation of the synchronized network is greater than that of the cell that had the lower fluctuation before the synchronization.

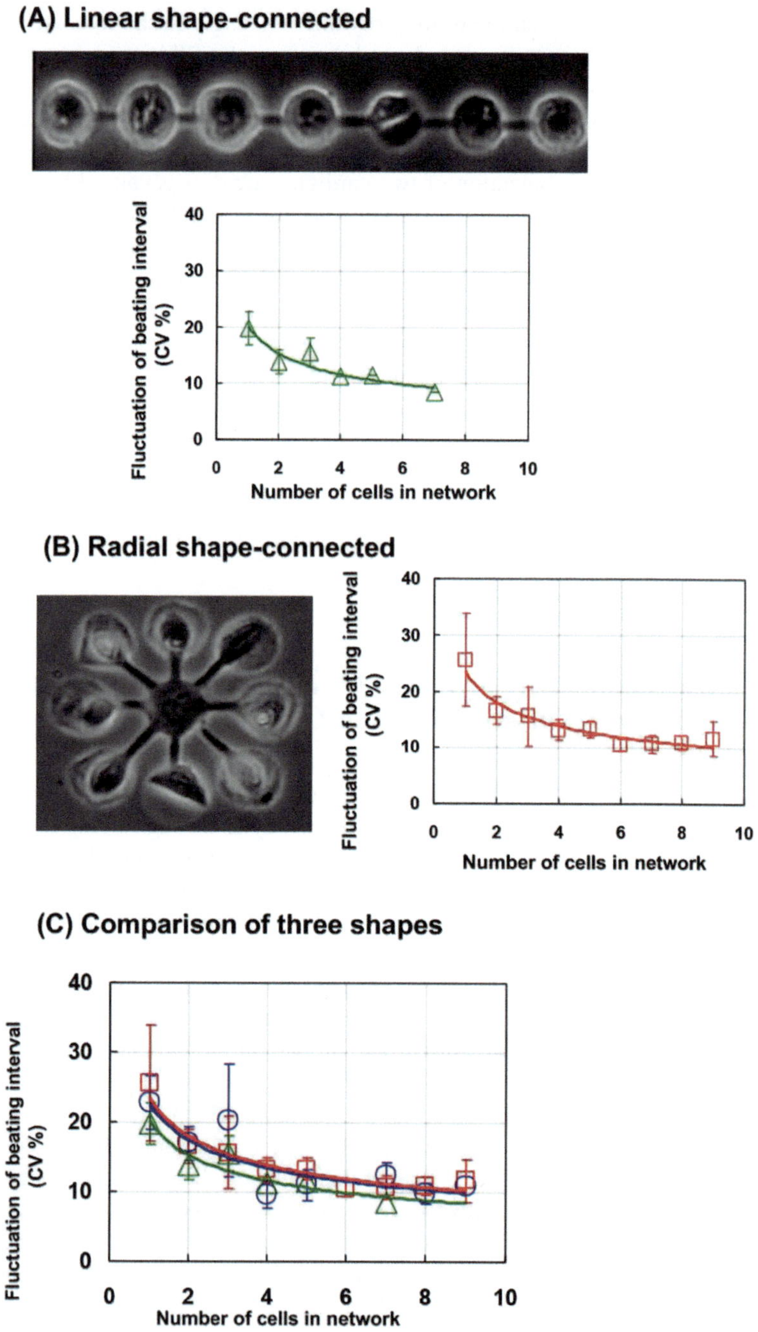

Fig. 3.15 Dependence of spatial arrangement of cardiomyocyte networks on cell number for beating stability. Three types of spatial arrangements: (**a**) linear shape, (**b**) radial shape, and lattice shape (see Fig. 3.14). (**c**) Fluctuation of beating interval versus network size in linear (green triangles), radial (red squares), and lattice (blue circles) cardiomyocyte networks. These plots show mean ± standard deviation

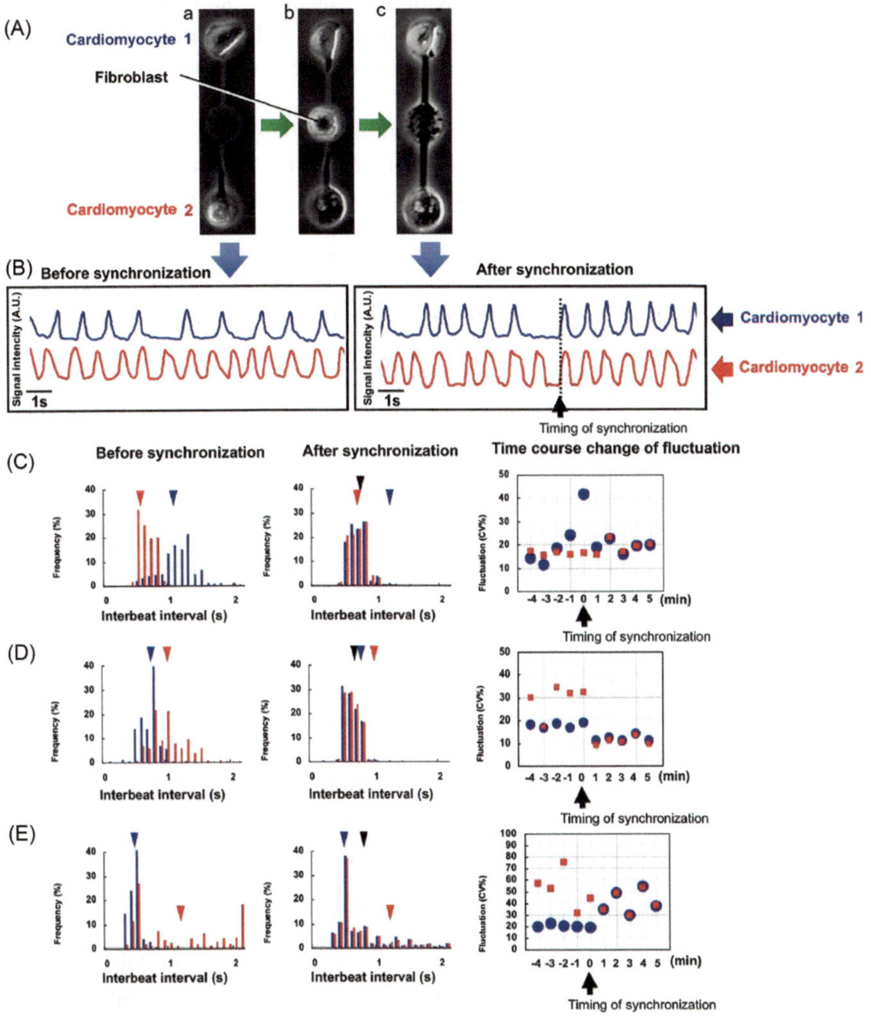

Fig. 3.16 Synchronization of two cardiomyocytes through a fibroblast. (**A**) Micrographs of two cardiomyocytes under isolated conditions (a), when a fibroblast was added between two cardiomyocytes (b), and when two cardiomyocytes were connected through a fibroblast and synchronization started (c). (**B**) Beating waveforms at (a) and (c) in (**A**). (**C**)–(**E**) Three types of synchronization tendencies. Beating frequency spectrum before (left graphs) and after (center graphs) synchronization and their beating fluctuation (right graphs). (**C**) Synchronization to a cell beating faster and more stably. (**D**) Synchronization and creation of new beating intervals contributing to beating stability. (**E**) Synchronization with new beating frequency, but beating fluctuation increased

Our photothermal etching method with agarose microchambers allows us to regulate the cell type and community size of cultured cells at the single-cell level. This could not be done when using the conventional cell cultivation method, so the prolific growth of cardiac fibroblasts made it difficult to culture only cardiomyocytes and investigate the properties of a single cell within a group of cells. By using single-cell-based cultivation, we were able to investigate how the fluctuation of the rates at which cardiomyocytes beat affects the reestablishment of synchronized beating.

3.3.6 Summary of Experimental Results

The results of the on-chip constructive experiments are summarized as follows:

1. When two isolated independently beating cardiomyocytes come into contact, they tend to beat synchronously at a rate that fluctuates no more than that of the cell whose beat rate fluctuated less than did that of the other cell.
2. When initially isolated cardiomyocytes form a network, its rhythm tends to entrain the beating rhythm of single cells whose beating rhythm fluctuated more than that of the network.
3. The entrainment activity of cell networks increases with their size, i.e., the fluctuation decreases.
4. Spatial arrangement does not affect the manner of synchronization of cardiomyocytes, and only the cell number of the network determines their tendency for synchronization.
5. The interbeat interval after the synchronization of two cardiomyocytes connected by a fibroblast is not the same as that after the synchronization of two cardiomyocytes directly connected to each other, and the tendency for the community effect to occur appears to be suppressed when the cardiomyocytes are heterogeneously coupled through a fibroblast.

They might indicate that unstable isolated cardiomyocytes reestablish a cell network that beats stably and synchronously. A novel finding of this study is that a cardiomyocyte network containing only a few cells acquires a stable rhythm. Moreover, once the cell or cell network achieves stable beating, an additionally attached unstable cell can synchronize to the stable cell or cell network and follow its stable beating intervals. This phenomenon also suggests that the factor of stability is very important in determining the fate of the beating frequency of the network after the connection of unstable cells.

3.3.7 *Ability and Limitation of Constructed Experimental Approach*

As described above, the constructed experimental approach is one of the potential solutions to solve the issue of quality control of cells. However, cells inherently display a variety of dynamic characteristics, even when cultivating cells in completely the same conditions and also when using those from the same single stem cells[33]. Figure 3.17 shows an example of this diversity of their expression. Although isolated single human iPS (hiPS) cardiomyocytes and human ES (hES) cardiomyocytes were derived from the same stem cells, their interbeat intervals (IBIs) and fluctuations [in this graph, we use the coefficient of variation (CV) as the index of fluctuation] were larger than those of the primary mouse cardiomyocytes. In contrast, two clusters made of the same dispersed cardiomyocytes showed similar

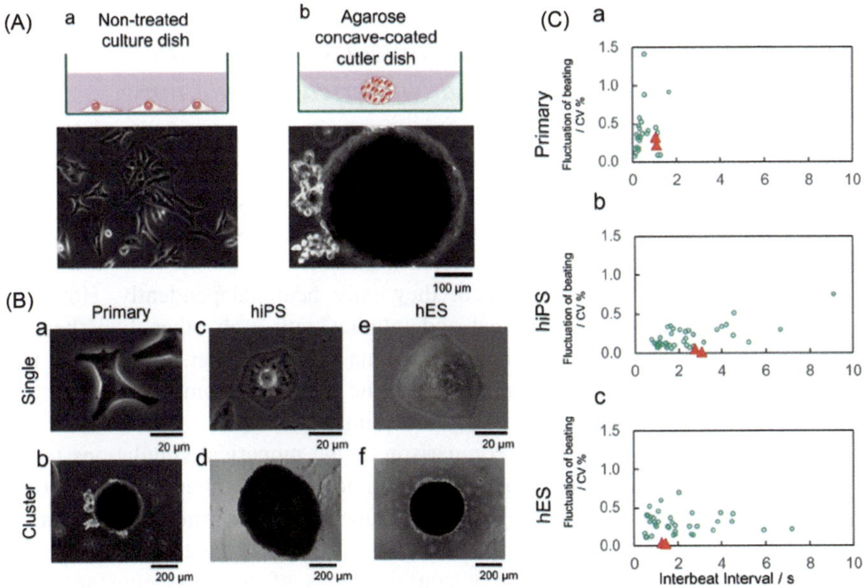

Fig. 3.17 Distribution of the interbeat intervals (IBIs) and fluctuations of isolated single cardiomyocytes and their clusters of primary, human iPS (hiPS), and human ES (hES) cells. (**A**) Method of cardiomyocyte cluster formation. (a) In the regular untreated culture dish, cardiomyocytes were dispersed and isolated. (b) In the agarose concave-coated culture dish, cardiomyocytes were gathered and clustered during incubation. (**B**) Micrographs of single cardiomyocytes and their clusters. (a) Isolated mouse primary cardiomyocyte, (b) clustered mouse primary cardiomyocytes, (c) single isolated hiPS cardiomyocyte, (d) clustered hiPS cardiomyocytes, (e) isolated hES cardiomyocyte, and (f) clustered hES cardiomyocytes. (**C**) Fluctuation–IBI relationship of mouse primary cardiomyocytes (a), hiPS cardiomyocytes (b), and hES cardiomyocytes (c). Green open circles are the 60-s mean IBIs and CVs of isolated single cardiomyocytes, and red filled triangles are those of clusters. The error bar indicates standard deviation. The coefficient of variation (CV) value of IBI was used as the fluctuation of beating

characteristics and appeared to represent their species (two red triangles in each graph of Fig. 3.17C). These results clearly indicate the ability and limitation of the experimental approach of single-cell-based assays. Each isolated single cell does not inherently show the same dynamics; however, once they formed a network, their diversity disappeared and stable shared characteristics appeared. We call this phenomenon the "community effect." To understand the meaning of the community effect, we need to have a set of completely controlled single cells. However, this is beyond the scope of the experimental approach. Even using the on-chip cellomics technologies, this experimental approach has a limitation of not allowing full control of the condition of all of the cells, especially in a dynamic context such as beating of the heart.

3.4 Numerical Approach to Synchronization of Cardiomyocytes

3.4.1 *Comparison of the Mathematical Modeling with Experimental Results and Numerical Simulations*

The mathematical model by modifying the integrate-and-fire model was shown in Sect. 3.2.1.1. This model is constructed on the basis of the simple Peskin's model [39], including refractory periods, stochastic process, and weak cell-to-cell interactions, which modulate phase variables [32, 56].

When cardiomyocytes are isolated, they only beat independently. However, if cardiomyocytes come into contact and interact with each other, their beating rhythms become synchronized. The experiments shown in Sect. 3.3 revealed that other cardiomyocytes are synchronized not to the fastest cardiomyocyte, but to the cardiomyocyte with the least fluctuation in beating rhythm [28].

In this section, we discuss the comparison of the numerical simulations to the experimental results and aim to theoretically understand the phenomenon that the synchronized beating of cardiomyocytes is tuned to the cardiomyocyte with a stable rhythm [13]. Using our model, we also aim to investigate the community effect of cardiomyocytes in different configurations of networks constituted by cardiomyocytes with specified characteristics of beating rhythms and to clarify how an assembly of cardiomyocytes acquires stability, one of the most important universal features in biological systems.

3.4.1.1 Mathematical Modeling for Synchronization of Cardiomyocytes

We consider a network of N cardiomyocytes and call ith cardiomyocyte cell i. The model was described by the phase variables $\phi_i(t)$ ($0 \leq \phi_i(t) \leq 2\pi$, $i = 1, 2, \ldots, N$), which denote the state of cell i at a time t. We assumed that the cell i fires (beats) when $\phi_i(t) = 0 (\equiv 2\pi)$. This firing occurs either at $\phi_i(t)$, reaches 2π, or the following conditions are satisfied: $\phi_i(t-0) \geq \theta_i$ ($\phi_i(t-0) :=$

$\lim_{\epsilon \to +0} \phi_i(t-\epsilon)$). Additionally, one of the cardiomyocytes connected to cell i (e.g., cell j) fired at a retardation time τ ago (i.e., $\phi_j(t-\tau) = 0$). Otherwise, we assumed that $\phi_i(t)$ is governed by the following interacting stochastic differential equation. Our mathematical modeling for cell i is as follows:

$$\begin{cases} d\phi_i(t) = \omega_i dt + dW(\sigma_i) + \sigma_i^2 \sum_j V(\phi_i, \phi_j)dt & (\phi_i(t-0) < \theta_i \text{ or } \phi_j(t-\tau) \neq 0), \\ \phi_i(t) = 0 & (\theta_i \leq \phi_i(t-0) \text{ and } \phi_j(t-\tau) = 0), \end{cases} \tag{3.49}$$

where ω_i is the average phase velocity of cell i, $dW(\sigma)$ is a stochastic process with standard deviation σ, and θ_i is a phase corresponding to the refractory period of cell $i(0 < \theta_i < 2\pi)$. $V(\phi_i, \phi_j)$ shows the weak interaction between cardiomyocytes through the membrane potential, which we assumed as the following form:

$$V(\phi_i, \phi_j) := \mu \sin(\phi_j - \phi_i), \tag{3.50}$$

where μ is a positive constant. An important point is that the stochastic process and the cell-to-cell interaction are correlated through the fluctuation–dissipation theorem that gives the relation between fluctuations and linear response to external force [31]. This will be discussed again in Sect. 3.4.3. The positive constant μ is the only free parameter in our model that cannot be directly determined by experiments, while ω_i, θ_i, and σ_i can be determined by single-cell experiments for each cardiomyocyte. In addition, we assumed that the boundary at $\phi_i(t) = 0$ is the reflective boundary condition, which ensures that the phase fluctuation is irreversible after firing. The schematic diagram of the dynamics of the phase variable $\phi_i(t)$ is shown in Fig. 3.18.

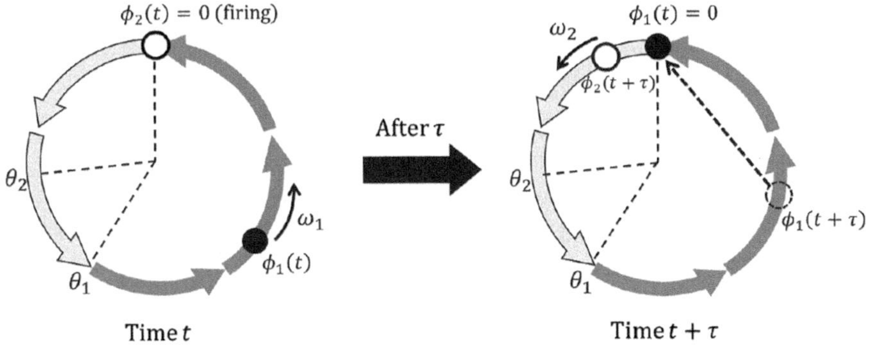

Fig. 3.18 Schematic diagram of the trajectory of the state variables $\phi_i(t)(i = 1, 2)$. The circle represents the trajectory of a state variable in the phase space of cardiomyocytes from one firing to the next firing. If cell 2 fires at a time t ($\phi_2(t)$=0) and cell 1 is not in the refractory period, then cell 1 fires at the retardation time τ after cell 2 fires

3.4.1.2 Numerical Simulation Method

The stochastic process in our simulation is described by an extended random walk. We used the following difference equations as a numerical approximation of Eq. (3.49). For almost all cardiomyocytes with a standard beating rhythm, we considered an ordinary random walk as follows:

$$\begin{cases} \phi_i(t+\Delta t) = \phi_i(t) + \omega_i \Delta t + \Delta\phi_i + \sigma_i^2 \sum_j V(\phi_i, \phi_j)\Delta t & (\phi_i(t) < \theta_i \text{ or } \phi_j(t-\tau) \neq 0), \\ \phi_i(t+\Delta t) = 0 & (\theta_i \leq \phi_i(t) \text{ and } \phi_j(t-\tau) = 0), \end{cases} \tag{3.51}$$

$$\Delta\phi_i = \begin{cases} +\Delta x_i & \text{(with probability 0.5)}, \\ -\Delta x_i & \text{(with probability 0.5)}, \end{cases} \tag{3.52}$$

where the standard deviation is defined by $\sigma := \Delta x/\sqrt{2\Delta t}$, Δt is the time difference interval, $\Delta x_i = \sqrt{2\Delta t\, \sigma_i^2}$ is the spatial difference determined by σ_i, and the delay time τ is set as $\Delta t \times k$ (k is a nonnegative integer). However, we could not reproduce the same beating fluctuation by using an ordinary random walk for cardiomyocytes with a large fluctuation. This is because the coefficient variation (CV%), which is defined by 100× standard deviation/mean beating rate, could be proved less than $100\sqrt{2/3} \simeq 81.65$ in Sect. 3.2.2. As shown in Fig. 3.13, some cardiomyocytes with the CV% which exceed this value are observed. Therefore, we adopted the following extended random walk, which is a history-dependent stochastic process, when beating fluctuation was larger than 81.65 (CV%):

$$\begin{cases} \phi_i(t+\Delta t) = \phi_i(t) + \omega_i \Delta t + \Delta\tilde{\phi}_i(t) + \sigma_i^2 \sum_j V(\phi_i, \phi_j)\Delta t & (\phi_i(t) < \theta_i \text{ or } \phi_j(t-\tau) \neq 0), \\ \phi_i(t+\Delta t) = 0 & (\theta_i \leq \phi_i(t) \text{ and } \phi_j(t-\tau) = 0). \end{cases} \tag{3.53}$$

The noise term $\Delta\tilde{\phi}_i(t)$ is defined as

$$\Delta\tilde{\phi}_i(t) = \begin{cases} +\Delta x_i & (\text{if } \Delta\tilde{\phi}_i(t-\Delta t) = \Delta x_i, \text{ then with probability } q), \\ 0 & (\text{if } \Delta\tilde{\phi}_i(t-\Delta t) = \Delta x_i, \text{ then with probability } 1-q), \\ 0 & (\text{if } \Delta\tilde{\phi}_i(t-\Delta t) = 0, \text{ then with probability } r), \\ +\Delta x_i & (\text{if } \Delta\tilde{\phi}_i(t-\Delta t) = 0, \text{ then with probability } 1-r). \end{cases} \tag{3.54}$$

However,

$$\Delta\tilde{\phi}_i(0) = \begin{cases} +\Delta x_i & \text{(with probability 0.5)}, \\ 0 & \text{(with probability 0.5)}. \end{cases} \tag{3.55}$$

The model could reproduce the large fluctuation observed in the experiments by setting appropriate values of q and r.

3.4.1.3 Comparison of the Model with Experimental Results of Two Cardiomyocytes

In the experiments shown in Sect. 3.3.3, the mean beating rate and its fluctuation before and after synchronization were observed for 14 pairs of cardiomyocytes (Fig. 3.13). We investigated whether our model could reproduce the results of these pairs of cardiomyocytes. We numbered these 14 pairs from Nos. 1 to 14 and distinguished the two cardiomyocytes in a pair by denoting "cell 1" and "cell 2." For each pair, we defined ω_i, σ_i, and θ_i in Eq. (3.49) for cell i ($i = 1, 2$), so that the model reproduced the same mean beating rate and fluctuation in beating rhythm. Since refractory periods of cardiomyocytes are almost the same as those for normal cardiomyocytes, we assumed that each cell had the common refractory period $t_{ref} = 0.3$ s. Therefore, θ_i is given by $\theta_i = t_{ref}\omega_i$. Figure 3.19 shows the mean beating rates and the beating fluctuation after synchronization for the 14 pairs obtained by the experiments and numerical results by our model. We could regard the retardation time τ as almost 0 because it was estimated as $10^{-3} \sim 10^{-4}$ of the mean beating rate. Therefore, we put $\tau = 0$. We used $\mu = 6.5$ in numerical

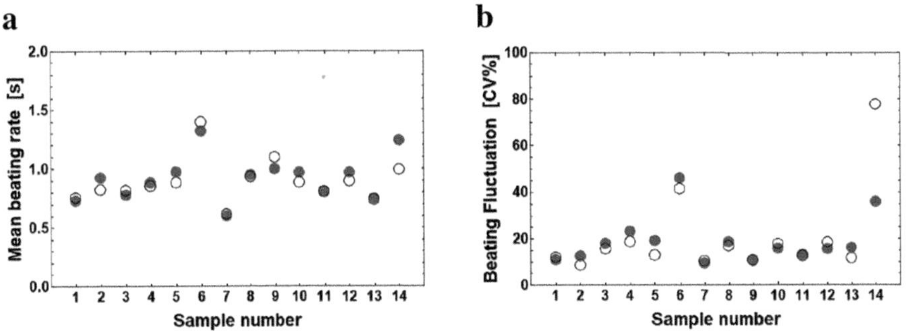

Fig. 3.19 The mean beating rate and beating fluctuation after synchronization. Numerical simulations for the 14 pairs of cardiomyocytes (28 cardiomyocytes) before synchronization in the experiments shown in Sect. 3.3.3 were performed using our integrate-and-fire model. Experimental values (circles) and theoretical values (filled circles) are plotted for (**a**) the mean beating rate and (**b**) beating fluctuation (CV%). For all numerical simulations, we used the same parameter values $\tau = 0$ and $\mu = 6.5$. The fluctuation in beating rhythm is expressed by the CV

simulations. The dependence of theoretical calculation on μ is shown later. We found that the simulated values accurately agree with the experimental values except for pair No. 14. The experimental result of pair No. 14 is exceptional because it is the only pair in which fluctuation increased after synchronization. Beating fluctuation of a pair of synchronized cardiomyocytes was equal to or less than that of less fluctuating cardiomyocytes, while the mean beating rate was widely distributed. Some pairs synchronized at faster rates of the two initial rates, some at slower rates of the two initial rates, and others at intermediate rates of the initial rates of the pair. We demonstrate how to determine the free parameter μ. Let us define an index ε, to evaluate the deviation of the theoretical values from the experimental values.

$$\varepsilon := \sum_{k=1}^{13}\sum_{i=1}^{2}\left\{(T_k^i - T_k^{\mathrm{ex}})^2 + (F_k^i - F_k^{\mathrm{ex}})^2\right\}, \tag{3.56}$$

where T_k^{ex} denotes the experimental value of the mean beating rate of the cell i in the pair k after synchronization, and $T_k^i (i = 1, 2)$ denotes the corresponding theoretical value obtained by our model. Similarly, F_k^{ex} denotes the experimental value of fluctuation (CV) of the cell i in the pair k after synchronization, and $F_k^i (i = 1, 2)$ denotes its theoretical value. Figure 3.20 shows the dependence of ε on μ. For $6 \lesssim \mu \lesssim 12$, ε keeps to take the lowest value. The results of the numerical simulations were almost constants for a relatively wide range of μ. This finding indicated that our model was robust against the free parameter μ.

3.4.1.4 Comparison with the Kuramoto Model

The two-oscillator phase model (the Kuramoto model [32]) with noise is as follows: for $i, j = 1, 2, i \neq j$,

$$d\psi_i(t) = \bar{\omega}_i dt + A_{i,j}\sin(\psi_j - \psi_i)dt + \bar{\sigma}_i dW_i(t), \qquad \psi_i(0) = 0, \tag{3.57}$$

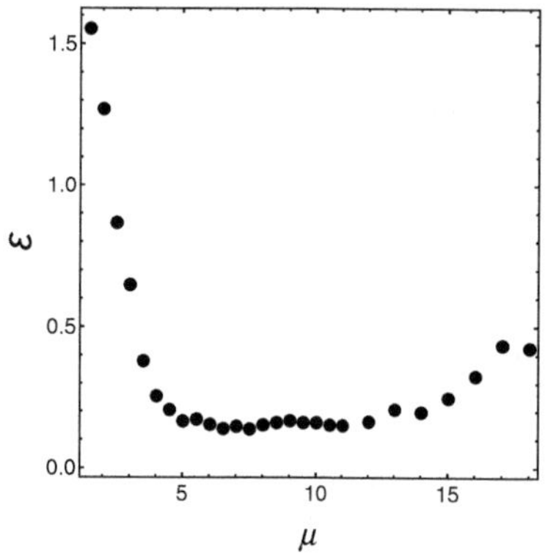

Fig. 3.20 Parameter μ dependence of deviation of numerical values from experimental data. The deviation between numerical simulation and experimental data is measured with the quantity ε by changing parameter μ

where $\bar{\omega_i}$ and $\bar{\sigma_i}$ denote the drift and noise strength constants, respectively, $A_{i,j}$ are nonnegative constants, and $\{W_i\}_{i=1,2}$ is independent standard Brownian motion (see Sect. 3.2.3). For two cases (Case (i) and Case (ii)), we applied the Kuramoto model (3.57) and our model (3.49) to synchronization of two coupled cardiomyocytes. The numerical simulation results were compared with biological experiment data (Fig. 3.12).

Case (i) A Case of Synchronization to a Cardiomyocyte with a Fast and Stable Beating Rhythm Two cardiomyocytes that we used in the Case (i) were cell 1 and cell 2 of pair No. 1, which have a mean beating rhythm of 0.64 s and fluctuation of 12.3 [CV%] and cell 2 with 1.23 s and 25.1 [CV%], respectively. When the two cardiomyocytes were coupled, we found that the bating rhythm after synchronization was tuned to cell 1 with a fast and stable beating rhythm (Fig. 3.21a). We investigated whether our model and the Kuramoto model could reproduce the experimental results. Figure 3.21b and c shows the theoretical predictions from our model and the Kuramoto model, respectively. The mean beating rate and beating fluctuation for the experimental result, our model, and the Kuramoto model are shown in Table 3.1.

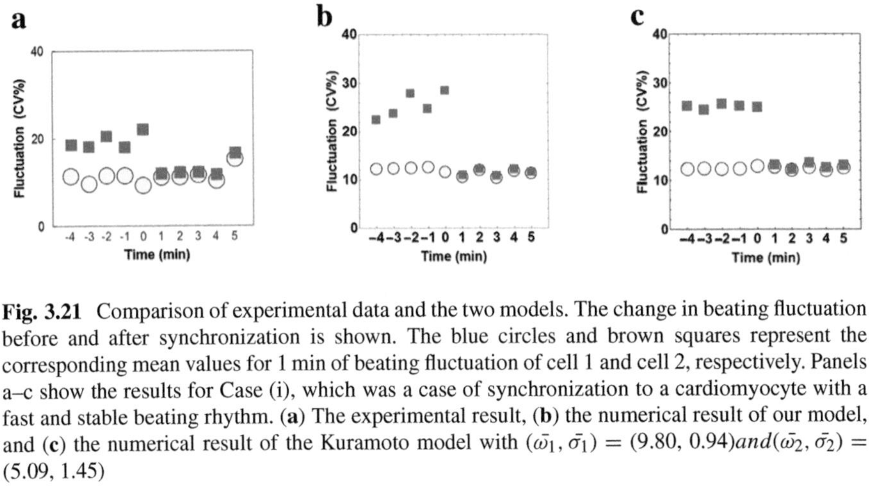

Fig. 3.21 Comparison of experimental data and the two models. The change in beating fluctuation before and after synchronization is shown. The blue circles and brown squares represent the corresponding mean values for 1 min of beating fluctuation of cell 1 and cell 2, respectively. Panels a–c show the results for Case (i), which was a case of synchronization to a cardiomyocyte with a fast and stable beating rhythm. (**a**) The experimental result, (**b**) the numerical result of our model, and (**c**) the numerical result of the Kuramoto model with $(\bar{\omega_1}, \bar{\sigma_1}) = (9.80, 0.94) and (\bar{\omega_2}, \bar{\sigma_2}) = (5.09, 1.45)$

Table 3.1 Comparison between the experimental result and the numerical results. The symbols T_i and F_i denote the mean beating rate and the beating fluctuation of the cell i $(i = 1, 2)$, respectively. The symbol T denotes the mean beating rate and F the beating fluctuation after synchronization

	Before synchronization				After synchronization	
	T_1(s)	F_1(CV%)	T_2(s)	F_2(CV%)	T(s)	F(CV%)
Experiments	0.64	12.3	1.23	25.1	0.76	12.3
Our model	0.64	12.3	1.23	25.1	0.74	11.4
Kuramoto model	0.64	12.3	1.23	25.1	0.85	12.7

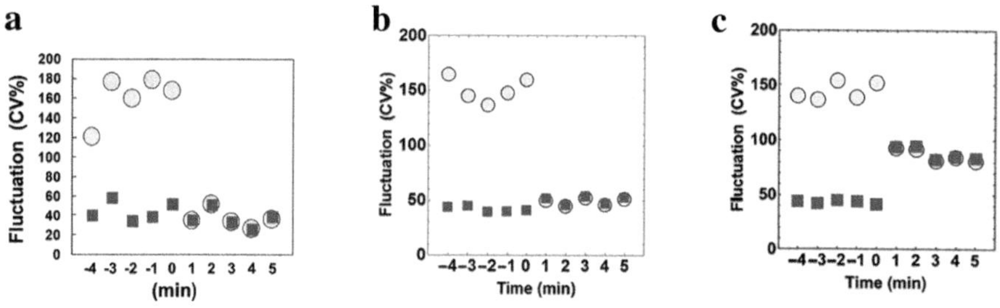

Fig. 3.22 Comparison of experimental data and the two models. The change in beating fluctuation before and after synchronization is shown. The blue circles and brown squares represent the corresponding mean values for 1 min of beating fluctuation of cell 1 and cell 2, respectively. Panels (**a**)–(**c**) show the results for Case (ii), which was a case of synchronization to a cardiomyocyte with a slow and stable beating rhythm. (**a**) The experimental result, (**b**) the numerical result of our model, and (**c**) the numerical result of the Kuramoto model with $(\bar{\omega_1}, \bar{\sigma_1}) = (5.03, 6.28) and (\bar{\omega_2}, \bar{\sigma_2}) = (4.46, 1.57)$

Table 3.2 Comparison between the experimental result and the numerical results. The symbols T_i and F_i denote the mean beating rate and the beating fluctuation of the cell $i\,(i = 1, 2)$, respectively. The symbol T denotes the mean beating rate and F the beating fluctuation after synchronization

	Before synchronization				After synchronization	
	T_1(s)	F_1(CV%)	T_2(s)	F_2(CV%)	T(s)	F(CV%)
Experiments	1.1	149	1.4	41.2	1.4	41.7
Our model	1.1	149	1.4	41.2	1.3	46.3
Kuramoto model	1.1	149	1.4	41.2	1.3	86.8

Case (ii) A Case of Synchronization to a Cardiomyocyte with a Slow and Stable Beating Rhythm Two cardiomyocytes that we used in the Case (ii) were cell 1 and cell 2 of pair No. 6, which have a mean beating rhythm of 1.10 s and fluctuation of 149 [CV%] and cell 2 with 1.40 s and 41.2 [CV%], respectively. When the two cardiomyocytes were coupled, we found that the bating rhythm after synchronization was tuned to cell 2 with a slow and stable beating rhythm (Fig. 3.22a). When we compared the numerical result of our model with that of the Kuramoto model, we found that our model was closer to the experimental data than the Kuramoto model. Our model showed that the beating rhythm after synchronization was tuned to the rhythm of the slow and stable cardiomyocyte (Fig. 3.22b). However, the Kuramoto model showed that beating fluctuation of the slow and stable cardiomyocyte was increased after synchronization, which differed from the experimental results (Fig. 3.22c). The mean beating rate and beating fluctuation of the experimental result, those of our model, and those of the Kuramoto model are shown in Table 3.2.

Therefore, our model showed that even though the mean beating rate of a cardiomyocyte was slow, a cardiomyocyte with more stable beating fluctuation dominated the beating rhythm after synchronization. In previous numerical simulations, we did not consider the effect of retardation time ($\tau = 0$). When we incorporated this effect, the behavior of our model barely changed because of the existence of a refractory period much longer than τ. However, if the refractory period is not taken into account, then a couple of cardiomyocytes continuously fire with the period of the retardation time, which is biologically unacceptable. In a system of two cardiomyocytes, we can use a retardation time $\tau = 0$, but we should consider the effect of retardation time as the size of the system increases. In this case, the existence of the refractory period will have significant effects on the system.

3.4.2 Numerical Experiments

As an application of our mathematical modeling, we then performed two numerical experiments on networks of cardiomyocytes and investigated the community effect of cardiomyocytes.

3.4.2.1 Size and Configuration Dependence on Fluctuation of the System

First, we investigated the dependence of fluctuation in beating rhythm of cardiomyocytes on the size and configuration of the system. Network patterns in cardiomyocyte groups that we considered were star, 2D lattice, and 1D lattice networks (Fig. 3.23).

We assumed that all the elements in cell networks have the same beating properties. Figure 3.24a–c shows the size dependence of fluctuation of networks with three types of configurations. The model cardiomyocyte that was used in Fig. 3.24a was cell 1 of pair No. 1, which had a mean beating rhythm of 0.64 s and fluctuation of 12.4 [CV%], that in Fig. 3.24b was cell 2 of pair No. 1 with

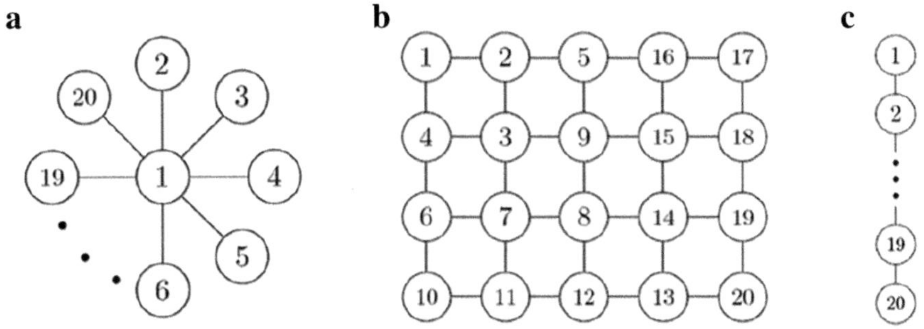

Fig. 3.23 Configurations and the order of placing cells. Three types of networks of configuration are shown. (**a**) Star network, (**b**) 2D lattice network, and (**c**) 1D lattice network. A cardiomyocyte is represented as a circle and it interacts with another cardiomyocyte if they are connected by a line. Cardiomyocytes are connected in ascending order according to the numbers in the circles from 1 to 20

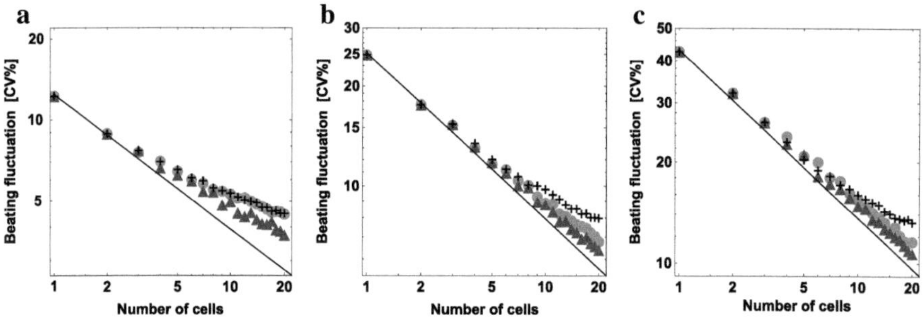

Fig. 3.24 Size dependence of fluctuation for three types of configuration. Size dependence of fluctuation is shown in double logarithmic graphs. The components of the network are model cardiomyocytes with the same characteristics. Panels(**a**)–(**c**) show the size dependence of fluctuation for three types of configurations: (**a**) $\omega = 9.80, \sigma = 0.69, \theta = 2.94$, (**b**) $\omega = 5.00, \sigma = 1.01, \theta = 1.50$, and (**c**) $\omega = 2.10, \sigma = 1.18, \theta = 0.63$. Circles indicate beating fluctuation (CV%) of cardiomyocytes in the star network, triangles indicate beating fluctuation in the 2D lattice network, and crosses indicate beating fluctuation in the 1D lattice network. The black straight line denotes $\propto N^{-1/2}$, where N is the number of cardiomyocytes in the network

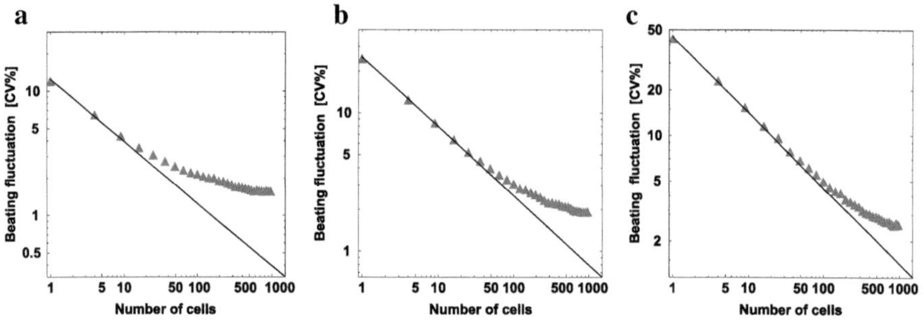

Fig. 3.25 Size dependence of fluctuation for a large network. Size dependence of fluctuation is shown in double logarithmic graphs. The components of the network are model cardiomyocytes with the same characteristics. Panels **a–c** show the size dependence of fluctuation for a larger 2D lattice network. (**a**) $\omega = 9.80, \sigma = 0.69, \theta = 2.94$, (**b**) $\omega = 5.00, \sigma = 1.01, \theta = 1.50$, and (**c**) $\omega = 2.10, \sigma = 1.18, \theta = 0.63$. The black straight line denotes $\propto N^{-1/2}$, where N is the number of cardiomyocytes in the network

1.23 s and 25.1 [CV%], and that in Fig. 3.24c was cell 2 of pair No. 8 with 2.71 s and 43.0 [CV%]. We found that the beating fluctuation decreased as the size of the system increased irrespective of network pattern. Among the three configurations, a reduction in fluctuation tended to be most rapid in the 2D lattice network, and fluctuation in the 1D lattice network tended to be always larger than that in the other two configurations. In addition, we considered the larger size (about 1000 cells) of the network in the 2D lattice network. Figure 3.25a–c shows the size dependence of fluctuation of the 2D lattice network where all the elements had the same beating properties. The numerical results suggested that the beating fluctuation decreased as the community size increased, but the CV value of the system approached a constant value for large system size N. For an ordinary stochastic ensemble, such as an independently identical distributed ensemble, the dependence of standard deviation of fluctuation on system size N was proportional to $N^{-1/2}$. However, we found that the data of fluctuation plotted on the graph (Fig. 3.25) considerably diverged from the line of $N^{-1/2}$ and the features of beating fluctuation behave differently from that of ordinary stochastic ensembles.

3.4.2.2 Dependence of Cell Properties and Numbers on Fluctuation of the System

We then investigated the change in beating rhythms after connecting two subsystems of cardiomyocytes. First, we prepared referential subsystems of four model cells and nine model cells. We assumed that these subsystems had the property of a standard beating rhythm (mean beating rate 1.20–1.30 s and fluctuation 15.0–20.0 [CV%]). As for the subsystems that are connected to referential subsystems, we considered subsystems comprising four types of cardiomyocytes: (1) first and stable cell, (2) first and unstable cell, (3) slow and stable cell, and (4) slow and unstable cell. We considered the three types of cell network patterns shown in Fig. 3.23. A single cardiomyocyte was connected to a center cardiomyocyte of the referential star network, to a cardiomyocyte on a link of the referential 2D lattice network, and to a cardiomyocyte on an edge of the 1D lattice network (Fig. 3.26a and b). When subsystems were the same size, we connected them by a single link between two cardiomyocytes at the same positions in the cell network. We used the two center cardiomyocytes for the star networks, cardiomyocytes on the links for the 2D lattice network, and the two cardiomyocytes at the edges for the 1D lattice network (Fig. 3.26c and d).

We showed three typical results of the numerical simulation. First, we considered the referential 2D lattice network with nine cardiomyocytes and a single cardiomyocyte with a fast and unstable beating rhythm (Fig. 3.27a). When the nine-cell network came into contact with the single cardiomyocyte, the single cardiomyocyte synchronized at the beating rhythm of the nine-cell network and with

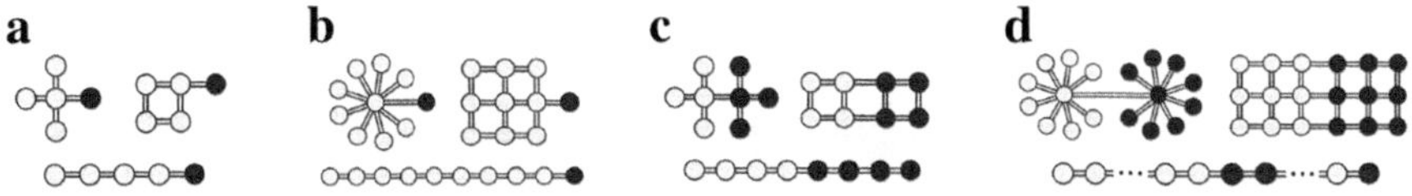

Fig. 3.26 Configurations of a combination between a referential network and a single cell or an assembly of cells. In panels (**a**) and (**b**), the filled circles denote a single cell, which adds to the referential network, of which cells are denoted by open circles. (**a**) Referential network of four cells + a single cell and (**b**) that of nine cells + a single cell. Panels (**c**) and (**d**) show the configurations of combined subsystems of four cells and those of nine cells. The cells in referential networks are denoted by open circles and those in counterparts are denoted by filled circles. (**c**) Referential network of four cells + four cells and (**d**) that of nine cells + nine cells

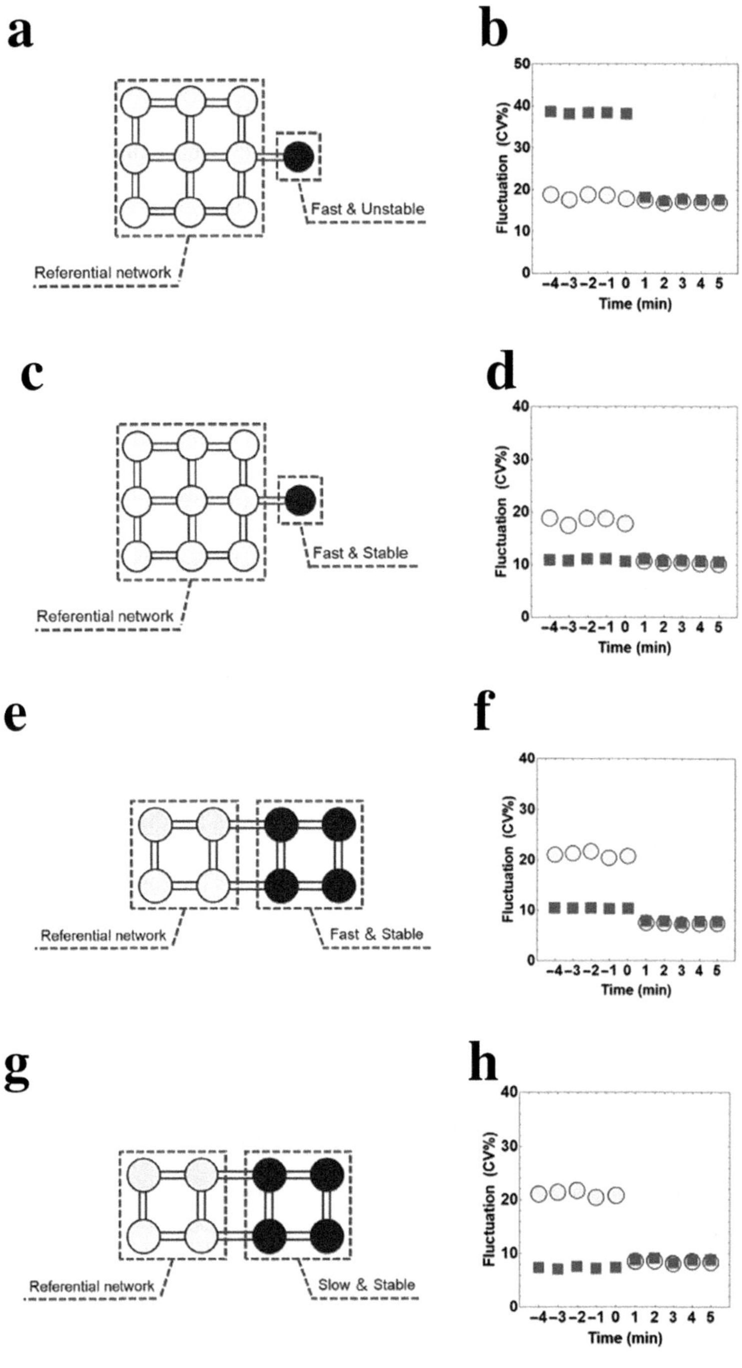

Fig. 3.27 Change in the beating fluctuation before and after synchronization. (**a**) The referential subsystem is the 2D lattice network and the counterpart is the single cell with a fast and (continued)

a beating fluctuation equal to that of the nine-cell network (Fig. 3.27b). Second, the rhythm of a single cardiomyocyte was fast and stable (Fig. 3.27c). We then found that even a single cardiomyocyte could lower fluctuation of the referential network and the beating rhythm of the referential network synchronized to a stable single cardiomyocyte (Fig. 3.27d). Finally, we considered a referential 2D lattice network with four cardiomyocytes and four cardiomyocytes grouped with a fast and stable beating rhythm (Fig. 3.27e). When the two subsystems were coupled, the synchronized beating rhythm was also tuned to the rhythm of the more stable group (Fig. 3.27f). In the above three cases (Fig. 3.27a–f), every cardiomyocyte started synchronizing when two subsystems were connected and formed a cell network. However, synchronization did not occur when the referential network was connected to a fast and stable 1D lattice network with nine cardiomyocytes. The cardiomyocytes near the edge of 1D lattice network with nine cardiomyocytes showed an exceptionally large fluctuation compared with the other cardiomyocytes. Furthermore, fluctuations of combined systems reduced their intensity, except for when there was a single cardiomyocyte or cardiomyocyte group with a slow and stable beating rhythm. However, the increment in fluctuation was small, even in these cases (e.g., Fig. 3.27g and h). When a referential subsystem came into contact with a counterpart consisting of one of the other three types of cardiomyocytes, the constituent cardiomyocytes acquired a common intensity of fluctuation. The intensity was intermediate between that of the prior two subsystems but was similar to that of the less fluctuating subsystem.

3.4.3 Discussion

To investigate the community effect of networks of cardiomyocytes, we used an interacting integrate-and-fire model with a refractory period. The reliability of the present mathematical model was verified by accurately reproducing recent experiments on pairs of cultured cardiomyocytes by Kojima et al.[28], despite the fact that the mathematical model has only one free parameter. One of the interesting points is that the fluctuation observed in their experiments cannot be accounted for simple Brownian motion or equivalently random walks. This is because some of the beating rhythm fluctuations (CV%) exceeded the theoretical limit evaluated for Brownian motion. An important observation in their experiments is the finding that when two isolated independently beating cardiomyocytes came

Fig. 3.27 (continued) unstable beating rhythm. (**b**) The change in mean value of beating fluctuation. The data for the referential networks and the counterparts are shown by circles and squares, respectively. The circles and squares show the corresponding mean values for 1 min of beating fluctuation. The results for the other combined systems (**c**), (**e**), and (**g**) are shown similarly in (**d**), (**f**), and (**h**), respectively

into contact, they tended to beat synchronously at a rate of the cardiomyocyte with a stable beating rhythm, but not the cardiomyocyte with a faster beating rhythm. This community effect of cardiomyocytes toward stability was confirmed with the present mathematical model by investigating cell networks of various configurations and constituent cardiomyocytes with various beating rhythms. Even a single stable cardiomyocyte could lower beating fluctuation of a cell network comprising some cardiomyocytes. The reason why a cardiomyocyte with an unstable beating rhythm tends to follow a cardiomyocyte with a stable beating rhythm may be explained as follows. A cardiomyocyte with a stable beating rhythm has the property where its dynamics are only slightly affected by external or internal disturbance. Therefore, there is little effect of interactions from neighboring cardiomyocytes. While, a cardiomyocyte with an unstable beating rhythm has the opposite property and is strongly affected by its neighbors. A cardiomyocyte with a stable beating rhythm corresponds to a pendulum with a heavy mass in contrast to a cardiomyocyte with an unstable beating rhythm that corresponds to that with a light mass (Fig. 3.28). When these pendulums are connected, the pendulum with a light mass tends to follow that with a heavy mass. This feature is a consequence of the fluctuation–dissipation theorem, which provides a universal relation between fluctuation and a linear response[3, 31]. In our model, the coefficient σ_i^2 of the interaction term $\sum_j V(\phi_j, \phi_i)$ in Eq. (3.49) was because of this theorem. This factor plays an essential role in stabilizing the beating rhythm after synchronization. Stability is one of the most significant and universal features of biological systems. It is an interesting finding that one of the origins of biological stability is a universal principle in statistical physics, that is, the fluctuation–dissipation theorem.

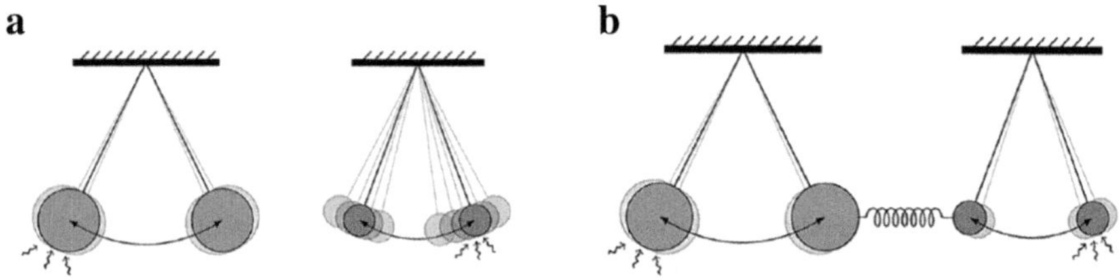

Fig. 3.28 Schematic explanation of why the beating rhythm tend to be synchronized to that of more stable cardiomyocytes after connection of two cardiomyocytes. A stable cardiomyocyte can be compared with a heavy pendulum and an unstable cardiomyocyte with a light pendulum. (**a**) External fluctuation has little effect on a pendulum's period of swing if its weight is heavy, but has strong effects if its weight is light. (**b**) When two pendulums are coupled and synchronized, their period of swing is close to that of the heavier pendulum, and fluctuation will be reduced because the total mass of weight increases

3.5 Summary

As described in this chapter, we examined the community effect with a mathematical model of cardiomyocyte synchronization behavior using the following three steps: first, to initiate the mathematical approach, we modeled the firing of cardiomyocytes as an oscillating stochastic phase model, involving the concepts of refractory period and induced beating, and also discussed its characteristics of single cardiomyocyte beating, two coupled cardiomyocytes, and the network of plurality of cardiomyocytes (Sect. 3.2). Next, we introduced the experimental results of the synchronous behavior of cardiomyocyte networks after a brief explanation of the experimental setup of microfabrication techniques regarding how the constructed approach of stepwise synchronization of cardiomyocytes was accomplished (Sect. 3.3). Finally, we proposed the oscillating stochastic phase model with the fluctuation–dissipation theorem and revealed that the model of cardiomyocyte networks with various cell numbers and spatial arrangements showed the same tendencies as the experimental results of synchronization behavior. Specifically, it revealed that the stability-oriented synchronization phenomenon and the fluctuation of beating intervals determine the cell network synchronous behavior (Sect. 3.4).

We here speculate about the macroscopic behavior behind the synchronization of beating cardiomyocytes. Such a synchronized network of living organisms appears to be a macroscopic system in which part of its behavior is not just purely mechanical, but it also exhibits statistical features that all systems tend to present. Hence, the community effect of cells should also be based on the statistical tendency of matter to become disordered as a part of the ordinary laws of physics.

At present, however, it is not clear whether and how this synchronization rule or community effect is regulated at the molecular ion channel level. In other words, no detailed information about the functioning of the community effect can emerge from a description of the genetic mechanism and its expression as general as that given above. In this regard, the next step for a mathematical approach to studying the community effect is to connect the macroscopic interpretation with the microscopic interpretation. For this, the in silico membrane potential model should become more precise [5, 7, 8, 16, 30, 38, 55], and it can also be applied in practical applications for drug discovery or predictive toxicity screening.

Finally, with regard to the community effect, living systems appear to maintain and perhaps encourage orderly and regulated behaviors, acting against the tendency for natural systems to progress from order to disorder, but based partly on some hidden existing order that is retained.

References

1. Anzai, Y., Terazono, H., Yasuda, K.: Simple non-invasive cell separation method using magnetic aptamer-conjugated microbeads and nuclease digestion. J. Biol. Phys. Chem. **7**, 83–86 (2007)
2. Burkitt, A.N.: A review of the integrate-and-fire neuron model: I. homogeneous synaptic input. Biolog. Cyber. **95**, 1–12 (2006)
3. Chandler, D.: Introduction to Modern Statistical Mechanics. Oxford University Press, Oxford (1987)
4. Çinlar, E.: Introduction to Stochastic Processes. Dover Publications, Mineola (2013)
5. Dokos, S., Celler, B., Lovell, N.: Ion currents underlying sinoatrial node pacemaker activity: a new single cell mathematical model. J. Theoret. Biol. **181**(3), 245–272 (1996). https://doi.org/10.1006/jtbi.1996.0129. http://linkinghub.elsevier.com/retrieve/pii/S0022519396901290
6. Evance, L.C.: An Introduction to Stochastic Differential Equations. American Mathematical Society, Providence (2013)
7. Fenton, F.H., Cherry, E.M., Hastings, H.M., Evans, S.J.: Real-time computer simulations of excitable media: java as a scientific language and as a wrapper for c and Fortran programs. Biosystems **64**(1–3), 73–96 (2002). https://doi.org/10.1016/S0303-2647(01)00177-0. http://linkinghub.elsevier.com/retrieve/pii/S0303264701001770
8. Hamada, H., Nomura, F., Kaneko, T., Yasuda, K., Okamoto, M.: Exploring the implicit interlayer regulatory mechanism between cells and tissue: Stochastic mathematical analyses of the spontaneous ordering in beating synchronization. Biosystems **111**(3), 208–215 (2013). https://doi.org/10.1016/j.biosystems.2013.02.007
9. Harrison, J.M.: Brownian Motion and Stochastic Flow Systems. Wiley, Hoboken (1985)
10. Hatano, A., Okada, J., Washio, T., Hisada, T., Sugiura, S.: A three-dimensional simulation model of cardiomyocyte integrating excitation-contraction coupling and metabolism. Biophys. J. **101**(11), 2601–2610 (2011)
11. Hattori, A., Moriguchi, H., Ishiwata, S., Yasuda, K.: A 1480/1064 nm dual wavelength photo-thermal etching system for non-contact three-dimensional microstructure generation into agar microculture chip. Sensors Actuators B Chem. **100**(3), 455–462 (2004). http://www.sciencedirect.com/science/article/B6THH-4C7DF4H-2/2/22c1b1c23921c20f35ec514510973c4b
12. Hayashi, M., Hattori, A., Kim, H., Terazono, H., Kaneko, T., Yasuda, K.: Fully automated on-chip imaging flow cytometry system with disposable contamination-free plastic re-cultivation chip. Int. J. Molec. Sci. **12**(6), 3618–3634 (2011). https://doi.org/10.3390/ijms12063618. http://www.mdpi.com/1422-0067/12/6/3618/http://www.ncbi.nlm.nih.gov/pubmed/21747698
13. Hayashi, T., Tokihiro, T., Kurihara, H., Yasuda, K.: Community effect of cardiomyocytes in beating rhythms is determined by stable cells. Sci. Rep. **7**, 15450 (2017)
14. Inoue, I., Wakamoto, Y., Yasuda, K.: Non-genetic variability of division cycle and growth of isolated individual cells in on-chip culture system. Proc. Jpn Acad. Ser. B Phys. Biolog. Sci. **77**(8), 145–150 (2001)
15. Inoue, I., Shiomi, D., Kawagishi, I., Yasuda, K.: Simultaneous measurement of sensor-protein dynamics and motility of a single cell by on-chip microcultivation system. J. Nanobiotechnol. **2**(1), 4 (2004). https://doi.org/10.1186/1477-3155-2-4. 1477-3155-2-4 [pii]. http://www.ncbi.nlm.nih.gov/pubmed/15119953
16. Joyner, R.W., Wilders, R., Wagner, M.B.: Propagation of pacemaker activity. Med. Biolog. Eng. Comput. **45**(2), 177–187 (2007). https://doi.org/10.1007/s11517-006-0102-9. http://link.springer.com/10.1007/s11517-006-0102-9
17. Kaneko, T., Kojima, K., Yasuda, K.: An on-chip cardiomyocyte cell network assay for stable drug screening regarding community effect of cell network size. Analyst **132**(9), 892–898 (2007). https://doi.org/10.1039/b704961g. http://www.ncbi.nlm.nih.gov/pubmed/17710264

18. Kaneko, T., Kojima, K., Yasuda, K.: Dependence of the community effect of cultured cardiomyocytes on the cell network pattern. Biochem. Biophys. Res. Commun. **356**(2), 494–498 (2007). https://doi.org/10.1016/j.bbrc.2007.03.005. http://www.ncbi.nlm.nih.gov/pubmed/17359936
19. Kaneko, T., Nomura, F., Yasuda, K.: On-chip constructive cell-network study (I): contribution of cardiac fibroblasts to cardiomyocyte beating synchronization and community effect. J. Nanobiotechnol. **9**(1), 21 (2011). https://doi.org/10.1186/1477-3155-9-21. http://www.ncbi.nlm.nih.gov/pubmed/21605419
20. Kaneko, T., Nomura, F., Hamada, T., Abe, Y., Takamori, H., Sakakura, T., Takasuna, K., Sanbuissho, A., Hyllner, J., Sartipy, P., Yasuda, K.: On-chip in vitro cell-network pre-clinical cardiac toxicity using spatiotemporal human cardiomyocyte measurement on a chip. Sci. Rep. **4**, 4670 (2014). https://doi.org/10.1038/srep04670. http://www.ncbi.nlm.nih.gov/pubmed/24751527. http://www.pubmedcentral.nih.gov/articlerender.fcgi?artid=PMC5381194
21. Kawai-Noma, S., Ayano, S., Pack, C.G., Kinjo, M., Yoshida, M., Yasuda, K., Taguchi, H.: Dynamics of yeast prion aggregates in single living cells. Genes Cells **11**(9), 1085–1096 (2006). https://doi.org/10.1111/j.1365-2443.2006.01004.x. http://www.ncbi.nlm.nih.gov/pubmed/16923127
22. Keener, J., Sneyd, J.: Mathematical Physiology. Springer, New York (1998)
23. Keener, J.P., Hoppensteadt, F.C., Rinzel, J.: Integrate-and-fire models of nerve membrane response to oscillatory input . SIAM J. Appl. Math. **41**, 503–517 (1981)
24. Kojima, K., Moriguchi, H., Hattori, A., Kaneko, T., Yasuda, K.: Two-dimensional network formation of cardiac myocytes in agar microculture chip with 1480 nm infrared laser photo-thermal etching. Lab Chip **3**(4), 292–296 (2003). https://doi.org/10.1039/b304652d. http://www.ncbi.nlm.nih.gov/pubmed/15007461
25. Kojima, K., Takahashi, K., Kaneko, T., Yasuda, K.: Flexible control of electrode pattern on cultivation chamber during cultivation of cells using nondestructive optical etching. Jpn J. Appl. Phys. **42**(Part 2, No. 8A), L980 (2003)
26. Kojima, K., Kaneko, T., Yasuda, K.: A novel method of cultivating cardiac myocytes in agarose microchamber chips for studying cell synchronization. J. Nanobiotechnol. **2**(1), 9 (2004). https://doi.org/10.1186/1477-3155-2-9. 1477-3155-2-9 [pii]. http://www.ncbi.nlm.nih.gov/pubmed/15357869
27. Kojima, K., Kaneko, T., Yasuda, K.: Stability of beating frequency in cardiac myocytes by their community effect measured by agarose microchamber chip. J. Nanobiotechnol. **3**(1), 4 (2005). https://doi.org/10.1186/1477-3155-3-4. http://www.jnanobiotechnology.com/content/3/1/4. http://www.ncbi.nlm.nih.gov/pubmed/15927047
28. Kojima, K., Kaneko, T., Yasuda, K.: Role of the community effect of cardiomyocyte in the entrainment and reestablishment of stable beating rhythms. Biochem. Biophys. Res. Commun. **351**(1), 209–215 (2006). https://doi.org/10.1016/j.bbrc.2006.10.037. http://www.ncbi.nlm.nih.gov/pubmed/17055457
29. Kori, K., Kawamura, Y., Masuda, N.: Structure of cell networks critically determines oscillation regularity . J. Theoret. Biol. **297**, 61–72 (2012)
30. Krogh-Madsen, T., Schaffer, P., Skriver, A.D., Taylor, L.K., Pelzmann, B., Koidl, B., Guevara, M.R.: An ionic model for rhythmic activity in small clusters of embryonic chick ventricular cells. Amer. J. Physiol.-Heart Circulat. Physiol. **289**(1), H398–H413 (2005). https://doi.org/10.1152/ajpheart.00683.2004. http://www.physiology.org/doi/10.1152/ajpheart.00683.2004
31. Kubo, R.: Statistical-mechanical theory of irreversible processes. i. general theory and simple applications to magnetic and conduction problems. J. Phys. Soc. Jpn **12**, 570–586 (1957)
32. Kuramoto, Y.: Chemical Oscillations, Waves, and Turbulence. Springer, New York (1984)
33. López-Redondo, F., Kurokawa, J., Nomura, F., Kaneko, T., Hamada, T., Furukawa, T., Yasuda, K.: A distribution analysis of action potential parameters obtained from patch-clamped human stem cell-derived cardiomyocytes. J. Pharmacolog. Sci. **131**(2), 141–145 (2016). https://doi.org/10.1016/j.jphs.2016.04.015. http://www.sciencedirect.com/science/article/pii/S1347861316300329

34. Matsumura, K., Yagi, T., Yasuda, K.: Role of timer and sizer in regulation of Chlamydomonas cell cycle. Biochem. Biophys. Res. Commun. **306**(4), 1042–1049 (2003). S0006291X03010891 [pii]. http://www.ncbi.nlm.nih.gov/pubmed/12821148
35. Mckean, H.P.: Stochastic Integrals. Academic, Cambridge (1969)
36. Moriguchi, H., Wakamoto, Y., Sugio, Y., Takahashi, K., Inoue, I., Yasuda, K.: An agar-microchamber cell-cultivation system: flexible change of microchamber shapes during cultivation by photo-thermal etching. Lab Chip **2**(2), 125–132 (2002). https://doi.org/10.1039/b202569h. http://www.ncbi.nlm.nih.gov/pubmed/15100846
37. Murray, J.: Mathematical Biology. Springer, Berlin (2002)
38. Noble, D.: Cardiac action and pacemaker potentials based on the Hodgkin-Huxley equations. Nature **188**(4749), 495–497 (1960). https://doi.org/10.1038/188495b0. http://www.nature.com/doifinder/10.1038/188495b0
39. Peskin, C.S.: Mathematical Aspects of Heart Physiology. Courant Institute of Mathematical Sciences, New York University (1975)
40. Sacerdote, L., Giraudo, M.T.: Stochastic integrable and fire models: A review on mathematical methods and their applications. In: Bachar M., Batzel J., Ditlevsen S. (eds.) Stochastic Biomathematical Models with Applications to Neuronal Modeling. Springer, Berlin (2013)
41. Skorokhod, A.V.: Stochastic equations for diffusion processes in a bounded region. Theory Probab. Appl. **6**, 264–274 (1961)
42. Spudich, J.L., Koshland, D.E.: Non-genetic individuality: chance in the single cell. Nature **262**(5568), 467–471 (1976). https://doi.org/10.1038/262467a0
43. Sugio, Y., Kojima, K., Moriguchi, H., Takahashi, K., Kaneko, T., Yasuda, K.: An agar-based on-chip neural-cell-cultivation system for stepwise control of network pattern generation during cultivation. Sensors Actuators B Chem. **99**(1), 156–162 (2004). http://www.sciencedirect.com/science/article/B6THH-49DFK5H3/2/8c54fed62b6b7e54b7452ad4cb40a836
44. Suzuki, I., Sugio, Y., Jimbo, Y., Yasuda, K.: Individual-cell-based electrophysiological measurement of a topographically controlled neuronal network pattern using agarose architecture with a multi-electrode array. Jpn J. Appl. Phys. **43**(3B), L403–L406 (2004)
45. Suzuki, I., Sugio, Y., Moriguchi, H., Jimbo, Y., Yasuda, K.: Modification of a neuronal network direction using stepwise photo-thermal etching of an agarose architecture. J. Nanobiotechnol. **2**(1), 7 (2004). https://doi.org/10.1186/1477-3155-2-7. 1477-3155-2-7 [pii]. http://www.ncbi.nlm.nih.gov/pubmed/15230976
46. Suzuki, I., Sugio, Y., Jimbo, Y., Yasuda, K.: Stepwise pattern modification of neuronal network in photo-thermally-etched agarose architecture on multi-electrode array chip for individual-cell-based electrophysiological measurement. Lab Chip **5**(3), 241–247 (2005). https://doi.org/10.1039/b406885h. http://www.ncbi.nlm.nih.gov/pubmed/15726199
47. Suzuki, I., Hattori, A., Yasuda, K.: On-chip multichannel action potential recording system for electrical measurement of single neurites of neuronal network. Jpn J. Appl. Phys. **46**(42), L1028–L1031 (2007)
48. Suzuki, I., Yasuda, K.: Constructive formation and connection of aligned micropatterned neural networks by stepwise photothermal etching during cultivation. Jpn J. Appl. Phys. **46**(9B), 6398–6403 (2007)
49. Suzuki, I., Yasuda, K.: Detection of tetanus-induced effects in linearly lined-up micropatterned neuronal networks: application of a multi-electrode array chip combined with agarose microstructures. Biochem. Biophys. Res. Commun. **356**(2), 470–475 (2007). https://doi.org/10.1016/j.bbrc.2007.03.006. http://www.ncbi.nlm.nih.gov/pubmed/17362877
50. Umehara, S., Hattori, A., Inoue, I., Yasuda, K.: Asynchrony in the growth and motility responses to environmental changes by individual bacterial cells. Biochem. Biophys. Res. Commun. **356**(2), 464–469 (2007). https://doi.org/10.1016/j.bbrc.2007.03.001. http://www.ncbi.nlm.nih.gov/pubmed/17350591
51. Umehara, S., Inoue, I., Wakamoto, Y., Yasuda, K.: Origin of individuality of two daughter cells during the division process examined by the simultaneous measurement of growth and swimming property using an on-chip single-cell cultivation system. Biophys. J. **93**(3), 1061–1067 (2007). https://doi.org/10.1529/biophysj.106.098061. http://www.ncbi.nlm.nih.gov/pubmed/17496044

52. Wakamoto, Y., Umehara, S., Matsumura, K., Inoue, I., Yasuda, K.: Development of non-destructive, non-contact single-cell based differential cell assay using on-chip microcultivation and optical tweezers. Sensors Actuators B Chem. **96**(3), 693–700 (2003). http://www.sciencedirect.com/science/article/B6THH-4979KM5-2/2/010b7f1a96b08e481751ecc3e0cb5f5f
53. Wakamoto, Y., Ramsden, J., Yasuda, K.: Single-cell growth and division dynamics showing epigenetic correlations. Analyst **130**(3), 311–317 (2005). https://doi.org/10.1039/b409860a. http://www.ncbi.nlm.nih.gov/pubmed/15724159
54. Wakamoto, Y., Yasuda, K.: Quantitative evaluation of cell-to-cell communication effects in cell group class using on-chip individual-cell-based cultivation system. Biochem. Biophys. Res. Commun. **349**(3), 1130–1138 (2006). https://doi.org/10.1016/j.bbrc.2006.08.149. http://www.ncbi.nlm.nih.gov/pubmed/16970916
55. Wilders, R.: Computer modelling of the sinoatrial node. Med. Biolog. Eng. Comput. **45** (2007). http://link.springer.com/10.1007/s11517-006-0147-9. http://link.springer.com/10.1007/s11517-006-0127-0
56. Winfree, A.: The Geometry of Biological Time. Springer, New York (2001)
57. Yasuda, K.: Non-destructive, non-contact handling method for biomaterials in micro-chamber by ultrasound. Sensors Actuators B Chem. **64**(1-3), 128–135 (2000). http://www.sciencedirect.com/science/article/B6THH-40BG617-S/2/a1469e7e1561311da3666e89cc25c1ca
58. Yasuda, K.: On-chip single-cell cultivation systems: Enabling algebraic and geometric understanding of cells. In: Andersson H., Berg A.v.d. (eds.) Lab-on-Chips for Cellomics, chap. Chapter 9, pp. 225–256. Kluwer Academic Publishers, Netherlands (2004)
59. Yasuda, K., Kiyama, M., Umemura, S.S.i., Takeda, K.: Deoxyribonucleic acid concentration using acoustic radiation force. J. Acoustical Soc. Amer. **99**(2), 1248–1251 (1996). http://link.aip.org/link/?JAS/99/1248/1
60. Yasuda, K., Umemura, S.i., Takeda, K.: Particle separation using acoustic radiation force and electrostatic force. J. Acoustical Soc. Amer. **99**(4), 1965–1970 (1996). http://link.aip.org/link/?JAS/99/1965/1
61. Yasuda, K., Haupt, S.S., Umemura, S.i., Yagi, T., Nishida, M., Shibata, Y.: Using acoustic radiation force as a concentration method for erythrocytes. J. Acoustical Soc. Amer. **102**(1), 642–645 (1997). http://link.aip.org/link/?JAS/102/642/1. http://www.ncbi.nlm.nih.gov/pubmed/9228824
62. Yasuda, K., Okano, K., Ishiwata, S.: Focal extraction of surface-bound DNA from a microchip using photo-thermal denaturation. Biotechniques **28**(5), 1006–1011 (2000). http://www.ncbi.nlm.nih.gov/pubmed/10818708

Chapter 4
Statistical Analysis of Cellular Directional Movement: Application for Research of Single Cell Movement

Masahiro Kanai, Kazuo Tonami, and Hideto Tozawa

4.1 Single Cell Movement

Cell movement plays a fundamental role in physiological collective phenomena, especially in self-organization. In earlier studies, it has been mainly discussed whether cell movement may be considered as random walk or not. We note that by *random walk* or *Brownian motion*, we denote the random motion of small particles driven by external force although these two terms have a different mathematical definition.

Due to Einstein's theory of Brownian motion, the mean squared displacement (MSD), which we can estimate from experimental data, is endorsed as a criterion for the motility to be random walk. The MSD of a cell trajectory $x(t)$, the position of a single cell at time t, is defined as the following equation:

$$\mathrm{MSD}(t) = \langle |x(t) - x(0)|^2 \rangle,$$

where $|x|$ and $\langle \cdots \rangle$ denote, respectively, the norm of vector x and the expectation. The MSD of a Brownian motion is proportional to time: MSD(t) $\propto t$, whereas that

M. Kanai (✉)
Kurume Institute of Technology, Kurume, Fukuoka, Japan
e-mail: kanai@kurume-it.ac.jp

K. Tonami
Graduate School of Medicine and Faculty of Medicine, The University of Tokyo, Bunkyo-ku, Tokyo, Japan
e-mail: kazutonami@g.ecc.u-tokyo.ac.jp

H. Tozawa
Department of Biological Sciences, Graduate School of Science, The University of Tokyo, Bunkyo-ku, Tokyo, Japan
e-mail: htozawa@g.ecc.u-tokyo.ac.jp

T. Tokihiro (ed.), *Mathematical Modeling for Genes to Collective Cell Dynamics*, Theoretical Biology, https://doi.org/10.1007/978-981-16-7132-6_4

of linear movement with a constant velocity is proportional to the square of time: MSD(t) $\propto t^2$. These two cases are called *diffusive* and *ballistic*, respectively, and the others are anomalous.

Recent studies [1, 2, 4, 6–9], based on experimental data of high accuracy, have tended to claim that cells move with a directional trend and a persistent random walk is suitable for modeling cell movement. The persistence parameter included in those models appears in the MSD and reveals a crossover from persistent movement to random. We will cover the persistent movement in the next section.

On the other hand, we consider another 2-dimensional model in which a cell moves with a constant speed prescribed and decides his direction entirely randomly at each time step. This model requires another method for analysis, namely circular statistics. In the latter section, as well as the MSD, circular statistics will be applied for the sample of cell movement.

4.1.1 Persistent Random Walk

One of the simplest ways to introduce *the persistent random walk* is to adopt, as equation of motion for cells, a Langevin-type equation in 2 dimension:

$$\tau \frac{dv}{dt} = -v + \sqrt{2D}\xi, \tag{4.1}$$

where $v = v(t)$ denotes the velocity vector of a cell at time t, and a vector $\xi = \xi(t) = \begin{pmatrix} \xi_1 \\ \xi_2 \end{pmatrix}$ represents the white noise, i.e., a random force which obeys the following properties:

$$\langle \xi \rangle = 0, \qquad \langle \xi_i(s)\xi_j(t) \rangle = \begin{cases} \delta(s-t) & (i = j) \\ 0 & (i \neq j) \end{cases},$$

where $\langle \cdots \rangle$ denotes the expectation. Here, $\delta(t)$ denotes Dirac's delta function which takes the value of 0 unless $t = 0$. The two components of the random force vector ξ are independent, and each has no auto-correlation. Note that one can arrange the coefficient of the friction term $-v$ to be 1 without loss of generality.

The parameter τ, taking positive values, does no longer represent the inertial mass because a cell can move by itself without any external force, and hence the relationship between the acceleration and the force is not retained in general. It is, however, still true that if τ is large the acceleration remains small. This means that the velocity will not change rapidly. Hence, τ should be regarded as a criterion for persistent movement and is called the persistence time since it has dimension of time.

The parameter D controls the strength of the random force and hence shows the motility of cells. From a macroscopic point of view, it corresponds exactly to the so-called diffusion coefficient originally introduced from Fick's law. Actually, in Fick's second law, the diffusion coefficient manifests itself in a diffusion equation.

4.1.2 Mean Squared Displacement (MSD) and the Fürth Formula

We calculate the MSD for the persistent random walk introduced in (4.1).

The velocity auto-correlation function (VACF) is obtained by integrating (4.1) as

$$\langle v(t) \cdot v(0) \rangle = \frac{2D}{\tau} e^{-|t|/\tau}, \tag{4.2}$$

where $v_1 \cdot v_2$ denotes the inner product of vectors v_1 and v_2, and $\langle \cdots \rangle$ denotes the expectation. The coefficient $\frac{2D}{\tau}$ is determined from the Green–Kubo formula:

$$D = \frac{1}{2} \int_0^\infty \langle v(t) \cdot v(0) \rangle dt.$$

In general, one can obtain the MSD from the VACF as follows.

Since

$$v = \frac{dx}{dt}$$

by definition, we have

$$\begin{aligned} \langle x(t)^2 \rangle &= \left\langle \left(\int_0^t v(s) ds \right)^2 \right\rangle \\ &= \left\langle \int_0^t v(s) ds \int_0^t v(s') ds' \right\rangle \\ &= \int_0^t \int_0^t \left\langle v(s) \cdot v(s') \right\rangle ds ds' \\ &= \int_0^t \int_0^t \left\langle v(s - s') \cdot v(0) \right\rangle ds ds'. \end{aligned} \tag{4.3}$$

Note that we can let $x(0) = (0, 0)$ without loss of generality. From (4.2), we calculate the integral and thus obtain the MSD referred to as the Fürth formula [3]:

$$\langle x(t)^2 \rangle = 4D(t - \tau(1 - e^{-t/\tau})). \tag{4.4}$$

We remark that, as already noted in [1, 8], the Fürth formula can be derived from other models than the Langevin equation.

In the short time region, i.e. for $t \ll \tau$, (4.4) is approximated as

$$\langle x(t)^2 \rangle \simeq \frac{2D}{\tau} t^2.$$

We note that from (4.2), the coefficient $\frac{2D}{\tau}$ coincides with $\langle v(0) \cdot v(0) \rangle$. This leads to $\sqrt{\langle x(t)^2 \rangle} \simeq \sqrt{\langle v(0) \cdot v(0) \rangle} t$, and hence this time region is called ballistic.

In the large time region, i.e. for $t \gg \tau$, (4.4) is approximated as

$$\langle x(t)^2 \rangle \simeq 4Dt.$$

This time region is called diffusive. In this region, the persistence time τ approximately vanishes in the MSD. This suggests that the mobility of cells may be considered as random walk in the long run.

4.2 Circular Statistics

In this section, we introduce statistical methods for analysis of *circular data*: numeric data measured in the form of angles or two-dimensional orientations.

Now, we have a wide variety of circular data as a set of vectors/axes, e.g., wind directions, circadian rhythms, cell division axis, directional movement of animals, and so on. In contrast, most scientists may not be familiar with the methods to deal with those data. The latter part of this section will be hence devoted to cell migration data measured as planar vectors; however, we do not refer to the general ways how these data should be recorded.

This section is concerned with basic methods for statistical analysis of a single sample of circular data $\{\theta_1, \theta_2, \ldots, \theta_n\}$ including methods for displaying and summarizing the sample. You may consider that the sample should present angles.

4.2.1 Raw Data Plot

First we consider the advantage of descriptive methods for statistical data. These enable us to gain an initial idea of the important characteristics of the sample: whether the sample does appear from a uniform distribution, from a unimodal distribution, or from a multimodal distribution. Moreover the distribution may be regarded as one of the fundamental distributions.

Here we refer to the *distribution* for a sample as the counts of sample points distributed over all possible values. Raw data plot is the first important step to analyzing the data because it implies that we will have the *next* measurement in the range in which the major part of the data concentrate.

Example: Wind Direction
Table 4.1 gives the sample of wind directions at *Kurume City, Fukuoka, Japan* observed on March 8th, 2018. The angles, figured in degree, are measured clockwise from the north direction. Figure 4.1 shows an angular plot for the raw data given in Table 4.1. This diagram enables one to recognize the whole aspect of the data.

4.2.2 Histograms

The next step is to exploit histograms. *Histograms* are constructed as a type of bar plot for numeric data that group the data into bins, i.e. a series of intervals dividing the entire range of values. First, there are two types of histograms/diagrams: linear and angular, and then some variations in the angular diagram. A table of unprocessed numerics is often referred to as raw data, compared to processed data (Fig. 4.2).

1. A simple angular histogram is obtained by plotting bars each of which is centered at the midpoint of its grouping interval, with the length of the bar proportional to the relative frequency in the group.
2. The *rose diagram* is more commonly used for angular data than the angular histogram above is, in which each group is displayed as a sector. The radius of each sector is taken so as to be proportional to the square root of the relative frequency of the group; the area of the sector is thus proportional to the group frequency (Fig. 4.3).

   Example: Wind Direction Statistics (Rose Diagram)
   Figure 4.4 shows a rose diagram for the data in Table 4.1. As immediately seen, the rose diagram is named for its shape.

3. The *stem-and-leaf* diagram is a histogram retaining the raw data values; in particular, their mode is immediately found from the sample. Each stem, as a column sequence of integers, consists of the data falling in a $10°$ interval with the leaves being the data values in that interval, sorted in an increasing order. The stem-and-leaf diagram is superior in displaying data without detracting the individual measurements.

   Example: Wind Direction Statistics (Stem–Leaf Diagram)
   Figure 4.3 gives a stem–leaf diagram for the data in Table 4.1. In order to make this diagram, we process the raw data in Table 4.1 by dividing all the numerics by $30°$. Accordingly, the stem consists of 12 items, and stems 0, 1, 2, 3, … present actual angles $0°,\ 30°, 60°,\ 90°,\ \ldots$, respectively. Then, for example, leaf $2\,|\,3$ present actual angle $67.5°$. We remark that this processing of the raw data changes numerics superficially but does not detract intrinsic statistics at all.

Table 4.1 Wind directions at *Kurume City, Fukuoka, Japan* on March 8, 2018 [5]

Time	Direction	Time	Direction	Time	Direction
0:10	45°	8:10	0°	16:10	22.5°
0:20	45°	8:20	337.5°	16:20	22.5°
0:30	45°	8:30	0°	16:30	0°
0:40	45°	8:40	337.5°	16:40	45°
0:50	45°	8:50	22.5°	16:50	22.5°
1:00	45°	9:00	0°	17:00	22.5°
1:10	22.5°	9:10	22.5°	17:10	22.5°
1:20	22.5°	9:20	0°	17:20	0°
1:30	45°	9:30	292.5°	17:30	22.5°
1:40	22.5°	9:40	45°	17:40	22.5°
1:50	22.5°	9:50	180°	17:50	22.5°
2:00	45°	10:00	225°	18:00	45°
2:10	45°	10:10	225°	18:10	22.5°
2:20	22.5°	10:20	157.5°	18:20	22.5°
2:30	22.5°	10:30	157.5°	18:30	0°
2:40	22.5°	10:40	157.5°	18:40	0°
2:50	22.5°	10:50	180°	18:50	45°
3:00	45°	11:00	135°	19:00	0°
3:10	45°	11:10	180°	19:10	0°
3:20	45°	11:20	180°	19:20	22.5°
3:30	45°	11:30	225°	19:30	22.5°
3:40	45°	11:40	202.5°	19:40	0°
3:50	45°	11:50	225°	19:50	0°
4:00	45°	12:00	247.5°	20:00	0°
4:10	45°	12:10	247.5°	20:10	22.5°
4:20	45°	12:20	247.5°	20:20	22.5°
4:30	45°	12:30	292.5°	20:30	22.5°
4:40	67.5°	12:40	270°	20:40	22.5°
4:50	22.5°	12:50	247.5°	20:50	0°
5:00	45°	13:00	247.5°	21:00	0°
5:10	22.5°	13:10	247.5°	21:10	22.5°
5:20	45°	13:20	247.5°	21:20	0°
5:30	45°	13:30	225°	21:30	0°
5:40	45°	13:40	225°	21:40	0°
5:50	0°	13:50	225°	21:50	22.5°
6:00	45°	14:00	225°	22:00	45°
6:10	22.5°	14:10	225°	22:10	22.5°
6:20	45°	14:20	225°	22:20	22.5°
6:30	45°	14:30	247.5°	22:30	22.5°
6:40	45°	14:40	225°	22:40	22.5°
6:50	45°	14:50	292.5°	22:50	22.5°

(continued)

Table 4.1 (continued)

Time	Direction	Time	Direction	Time	Direction
7:00	67.5°	15:00	337.5°	23:00	22.5°
7:10	45°	15:10	0°	23:10	22.5°
7:20	67.5°	15:20	90°	23:20	337.5°
7:30	45°	15:30	22.5°	23:30	0°
7:40	22.5°	15:40	0°	23:40	0°
7:50	0°	15:50	0°	23:50	0°
8:00	22.5°	16:00	22.5°	0:00	0°

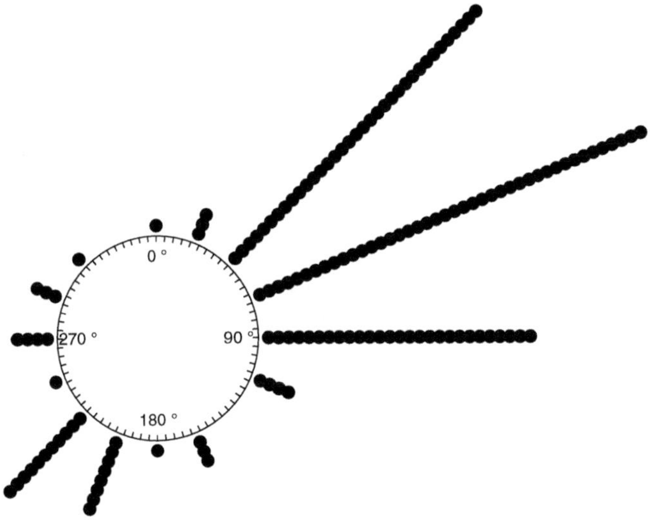

Fig. 4.1 An angular plot for the data on the raw data is given in Table 4.1. The angles are measured clockwise from the north direction

4. We introduce a *kernel density estimation*, which is one of nonparametric methods to estimate the distribution for a sample. This method is different from the histograms described above in that the contribution of each data point θ_k is defined as

$$\frac{1}{nh} w\left(\frac{\theta - \theta_k}{h}\right) d\theta,$$

where $w(\theta)$ presents a smooth function of θ taking the form of a bump, $d\theta$ denotes an infinitely small interval, and h controls the amount of smoothing effect. Then, the kernel density estimation $\hat{f}(\theta)$ is given by the sum of all the contributions:

$$\hat{f}(\theta) d\theta = \frac{1}{nh} \sum_{k=1}^{n} w\left(\frac{\theta - \theta_k}{h}\right) d\theta.$$

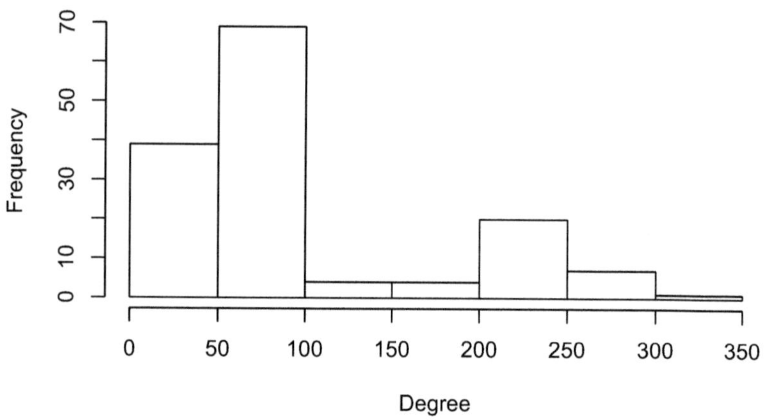

Fig. 4.2 A linear histogram for the data on the wind directions at *Kurume City, Fukuoka, Japan*. The binwidth is 50°. The raw data is given in Table 4.1

0	0 8 8 8
1	5 5 5 5 5 5 5 5 5 5 5 5 5 5 5 5 5 5 5 5 5 5 5 5 5 5 5 5 5 5 5 5 5 5 5
2	3 3 3 3 3 3 3 3 3 3 3 3 3 3 3 3 3 3 3 3 3 3 3 3 3 3 3 3 3 3 3 3 3 3 3 3 3 3 3 3 3 3
3	0 0 0 0 0 0 0 0 0 0 0 0 0 0 0 0 0 0 0 0 0 0 0 0 0 0 0 8 8 8 8
4	
5	3 3 3
6	0 8 8 8 8 8 8 8 8
7	5 5 5 5 5 5 5 5 5 5 5
8	3
9	0 0 0 0 8 8 8
10	5
11	

Fig. 4.3 A stem–leaf diagram for the wind directions at *Kurume City, Fukuoka, Japan* on March 8, 2018 [5]

Function $w(\theta)$ called the kernel function should be chosen so that its support D, a region where the function takes nonzero values, is small enough for $w(\theta)$ to ensure the condition

$$\int_D w(\theta)d\theta = 1.$$

However, the choice is not crucial to the density estimation but the magnitude of h is because it determines the amount of smoothing effect, namely, the larger h is, the more the density function gets blurred. It is, however, another issue how you determine the value of h.

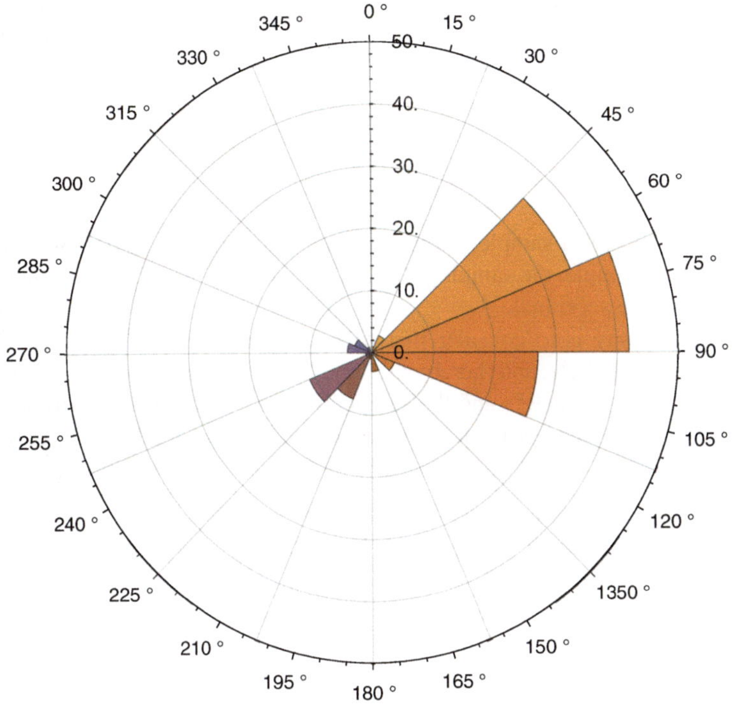

Fig. 4.4 A rose diagram for the data on the wind directions at *Kurume City, Fukuoka, Japan*. The raw data is given in Table 4.1

We note some important points in plotting histograms: Some choice of group boundaries can give rise to a serious distortion of the information about the modal groups observed in the sample; as well, the choice of the bin width should be carefully made so that we can anticipate the shape of the underlying distribution of the sample. It is, however, another issue whether your choice is correct or not.

4.2.3 Summary Statistics

The diagrams described above suggest the existence of the population from which the sample was drawn. Here we introduce basic quantities which describe important features of the sample distribution.

(A) Sample Circular Mean and Sample Circular Standard Deviation
The most popular as well as important statistic is the sample mean, which is ordinarily given by the arithmetic mean:

$$\frac{1}{n}\sum_{j=1}^{n}\theta_j.$$

Is it correct even for a sample of circular data? No, and it will be obvious if one considers, for example, the sample $\{10°, 350°\}$. We have the arithmetic mean 180°, but this cannot be acceptable.

The natural way is to consider that each circular data corresponds to a point located on the unit circle, and thereby the sample of circular data directly transforms into a sample of unit vectors and the sample mean for circular data is to be figured out from the mean vector. The mean vector in terms of vector addition is carried out as

$$C = \frac{1}{n}\sum_{j=1}^{n}\cos\theta_j, \qquad S = \frac{1}{n}\sum_{j=1}^{n}\sin\theta_j, \qquad R = \sqrt{C^2 + S^2},$$

where R should be chosen to be positive. Then, the mean direction $\bar{\theta}$ of the sample of unit vectors is obtained from

$$\cos\bar{\theta} = \frac{C}{R}, \qquad \sin\bar{\theta} = \frac{S}{R}.$$

This mean direction $\bar{\theta}$ corresponds to the mean of the sample of circular data θ_is.

On the other hand, R defined above presents the length of the resultant vector and is no longer a unit vector. R takes the value in the range [0, 1]; if $R = 1$, it means that all the vectors are in the same direction and hence all data θ_is are coincident. By contrast, $R = 0$ does not always mean that all the unit vectors distribute in uniformly random directions. A simple counterexample is the sample data $\{10°, 180°, 350°\}$. This point will become more clear when we consider statistics for axial data. We hence note that R cannot be a useful measure of deviation.

However, R may present the variance as

$$V = 1 - R$$

in the case that the sample shows a single modal distribution. We note that circular data is restricted to a finite range and so is the variance. Again, $V = 1$ does not immediately imply that the sample has a dispersed distribution.

The sample circular standard deviation is also defined as

$$v = \sqrt{-2\log(1 - V)}.$$

(We may have other definitions for the standard deviation.) If V is nearly equal to 1, v can be well approximated by $\sqrt{2V}$.

(B) Advanced Circular Summary Statistics

In order to define advanced statistics, we need some mathematics in complex numbers. From the mean direction $\bar{\theta}$ and the resultant vector length R, we define the first trigonometric moment

$$m_1 = R(\cos\bar{\theta} + i \sin\bar{\theta}),$$

where $i = \sqrt{-1}$. In an analogous manner as the mean direction, the pth trigonometric moment is then defined as

$$m_p = R_p(C_p + i S_p),$$

where

$$C_p = \frac{1}{n}\sum_{i=1}^{n} \cos p\theta_i, \qquad S_p = \frac{1}{n}\sum_{i=1}^{n} \sin p\theta_i, \qquad R_p = \sqrt{C_p^2 + S_p^2}.$$

Note that $R_1 = R$.

Using Euler's formula in complex analysis, $e^{i\theta} = \cos\theta + i \sin\theta$, we have a simple expression:

$$C_p + i S_p = \frac{1}{n}\sum_{j=1}^{n} e^{ip\theta_j}.$$

Then, we introduce argument μ_p as

$$R_p e^{i\mu_p} = \frac{1}{n}\sum_{j=1}^{n} e^{ip\theta_j}.$$

In particular, $\mu_1 = \bar{\theta}$, but however $\mu_p \neq p\bar{\theta}$ for $p \geq 2$ in general.

The centered sample trigonometric moments are also defined as

$$m'_p = R'_p(C'_p + i S'_p),$$

where

$$C'_p = \frac{1}{n}\sum_{j=1}^{n} \cos p(\theta_j - \bar{\theta}), \qquad S'_p = \frac{1}{n}\sum_{j=1}^{n} \sin p(\theta_j - \bar{\theta}),$$

and

$$R'_p = \sqrt{(C'_p)^2 + (S'_p)^2}.$$

Since a little calculation leads to $C'_1 = R$ and $S'_1 = 0$, we have $m'_1 = R$.

Using Euler's formula, we have

$$R'_p e^{i\mu'_p} = \frac{1}{n}\sum_{j=1}^{n} e^{ip(\theta_j - \bar{\theta})} = \frac{e^{-ip\bar{\theta}}}{n}\sum_{j=1}^{n} e^{ip\theta_j} = R_p e^{i(\mu_p - p\bar{\theta})},$$

and therefore $\mu'_p = \mu_p - p\bar{\theta}$, and $R'_p = R_p$.

As for a unimodal distribution of the sample, we exploit the first and second centered trigonometric moments, defining advanced statistics: the sample circular dispersion

$$\delta = \frac{1 - R'_2}{2(R'_1)^2},$$

the sample circular skewness

$$s = \frac{R'_2 \sin(\mu'_2 - 2\mu'_1)}{\sqrt[3]{1 - R'_1}},$$

and the sample circular kurtosis

$$k = \frac{R'_2 \cos(\mu'_2 - 2\mu'_1) - (R'_1)^4}{(1 - R'_1)^2}.$$

The circular dispersion δ is concerned with confidence interval for the sample mean direction. The skewness s presents the asymmetry of the sample distribution. The kurtosis k presents the peakedness of the sample distribution, and so it is also called peakedness.

Example: Wind Direction Statistics

We have summary quantities for the wind directions given in Table 4.1 as follows: the mean direction $\bar{\theta} = 71.7°$, the sample circular variance $V = 0.43$, and the circular standard deviation $v = 1.1$. See Figs. 4.2 and 4.4, and verify these results.

4.2.4 Probability Models

Statistical analysis, especially statistical inference, of a sample data is based on the probability of it being obtained. Probability in statistics takes a role in formulating uncertainty of the data, i.e., the data we obtain is drawn from an underlying population. Each sample of the population contains the probability that one will draw it, and we call the set of the probability the probability distribution or distribution simply.

Probability models provide with a distribution for the data a priori. Some probability models have parameters to be inferred from the sample, and other models called nonparametric do not. One usually applies the normal distribution for linear data, then obtaining acceptable results. For circular data, some models appear as a counterpart of the normal distribution for linear data, and the von Mises distribution described below is one of them.

Example: von Mises Distribution

Figure 4.5 shows a von Mises distribution. This distribution is a continuous probability distribution on the circle and was introduced as a circular analogue of the normal distribution. The probability density function is defined by

$$f(\theta) = \frac{e^{\kappa \cos(\theta-\mu)}}{2\pi I_0(\kappa)} \qquad (0 \leq \theta < 2\pi,\ 0 \leq \kappa < \infty),$$

where the parameters μ and $1/\kappa$ correspond, respectively, to the mean value and the variance. We note that the normalization is given by a modified Bessel function $I_0(\kappa)$. We have some analytic expressions for the function as

$$I_0(\kappa) = \frac{1}{2\pi} \int_0^{2\pi} e^{\kappa \cos(\theta-\mu)} d\theta,$$

and

$$I_0(\kappa) = \sum_{r=0}^{\infty} (r!)^{-2} \left(\frac{\kappa}{2}\right)^{2r}.$$

However, these may not be useful for numerical calculation. The use of a software on computer is practical.

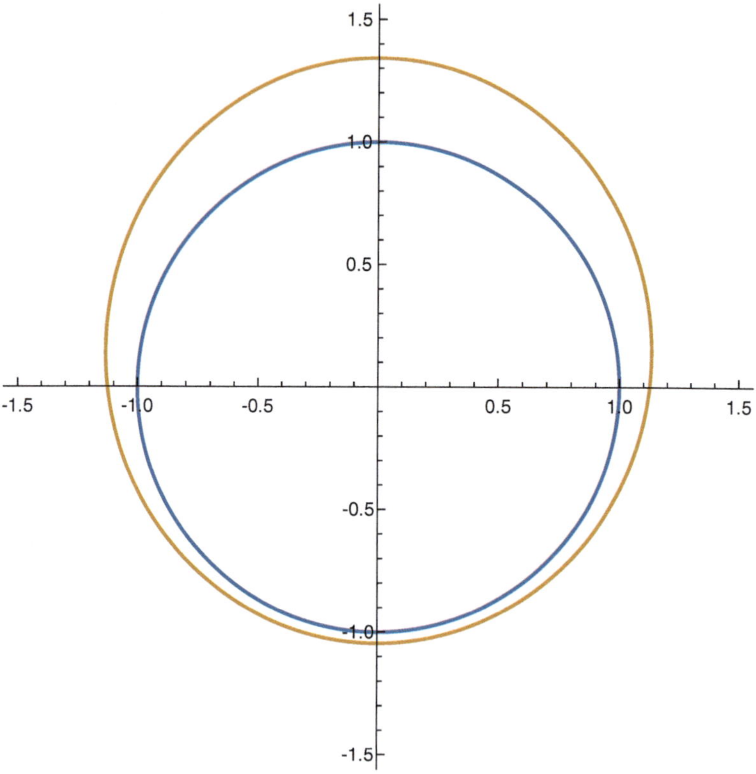

Fig. 4.5 The von Mises distribution. We illustrate a von Mises distribution in polar plot; the outer curve presents the value of probability increasing upon the unit circle (the inner curve)

For example, we apply the von Mises distribution to the wind direction data given in Table 4.1. As already calculated above, the circular mean $\mu = 71.7°$ and the circular variance $1/\kappa = 0.43$. Figure 4.6 shows the probability density function in polar plot.

4.3 Application for Single Cell Movement

In this final section, we apply the circular statistical analysis introduced above for single cell movement. We obtain the sample of a single 3T3 cell in vitro. The 3T3 cell line is a spontaneously immortalized mouse fibroblast cell line established from mouse embryonic tissue.

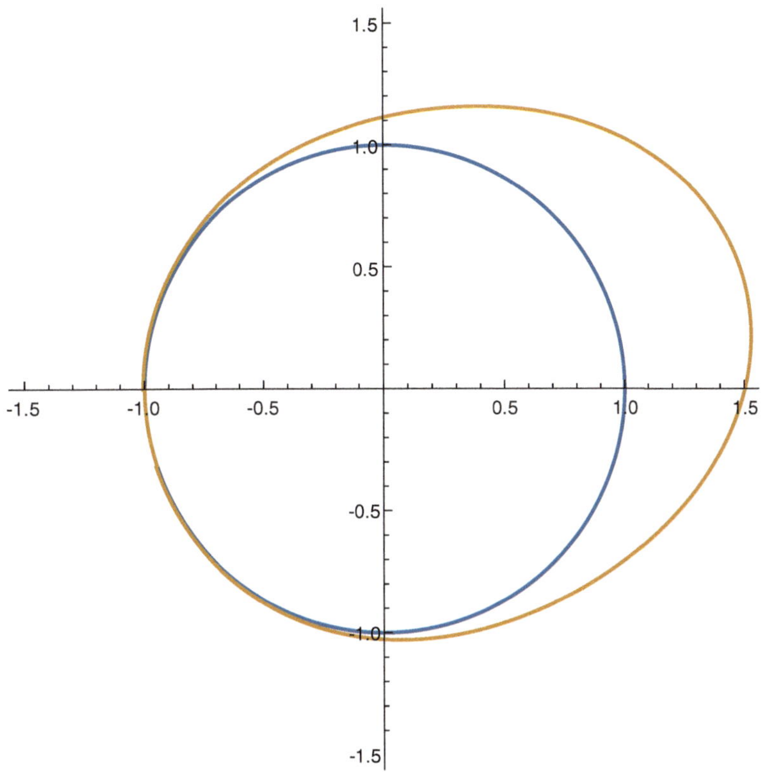

Fig. 4.6 The von Mises distribution applied for the wind direction data given in Table 4.1. Parameters are chosen so that the mean $\mu = 71.7°$ and the variance $1/\kappa = 0.43$. We find that this fits with the rose diagram given in Fig. 4.4. To be precise, we illustrate the graph of polar equation $\rho = f(\theta) + 1$ with $\mu = 1.25$ and the variance $\kappa = 2.33$

Figure 4.7 shows the sample trajectory of a single 3T3 cell moving on a dish freely. We plot the data points by time lapse imaging of the cell movement and then connect them with line segments in time sequence.

Figure 4.8 shows the directions of motion obtained from the sample trajectory in both circular histogram and rose diagram. From the data, we have the mean circular mean direction 121° and the circular variance 0.93. Since the circular variance takes a value in between 0 and 1 and the greater it is the more random (irregular) the movement is, we consider from the present result that the 3T3 cell chooses the direction of movement with almost equal probability (Fig. 4.9).

Furthermore, we calculate the mean squared displacement from the sample trajectory data, so we test that the movement observed is a random walk or not. We use a log–log plot for the figure, and hence the line of the linear minimum mean squared error estimator gives the exponent λ for the MSD with respect to time: MSD $\propto t^{\lambda}$. The slope of the line is read as 1.0, i.e. $\lambda = 1.0$, and accordingly the result suggests that the present cell movement should be considered as diffusive. Hence

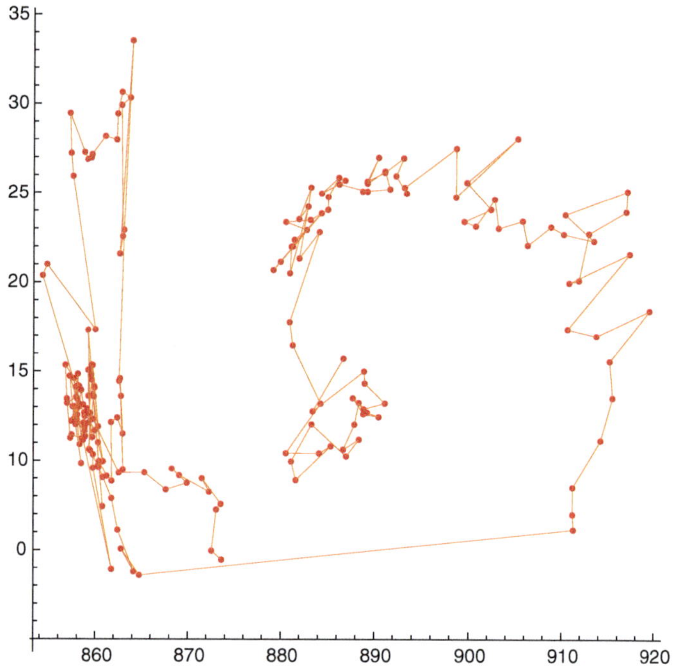

Fig. 4.7 The sample of a single 3T3 cell moving freely on a dish

we conclude from the sample data given in Fig. 4.7 that the 3T3 cell moves as a random walk.

We finally make a remark on the calculation of the MSD from the sample data before closing the chapter. From a sample trajectory containing $N+1$ points of position $(x(t), y(t))$ at time t, we actually compute the MSD as

$$\begin{aligned}\mathrm{MSD}(\tau) &= \langle (x(t+\tau)-x(t))^2 + (y(t+\tau)-y(t))^2 \rangle \\ &= \frac{1}{N-n+1}\sum_{i=0}^{N-n}\Big[(x((i+n)\Delta t)-x(i\Delta t))^2 \\ &\quad +(y((i+n)\Delta t)-y(i\Delta t))^2\Big],\end{aligned} \tag{4.5}$$

where $\tau = n\Delta t$ $(n = 0, 1, 2, \ldots, N)$.

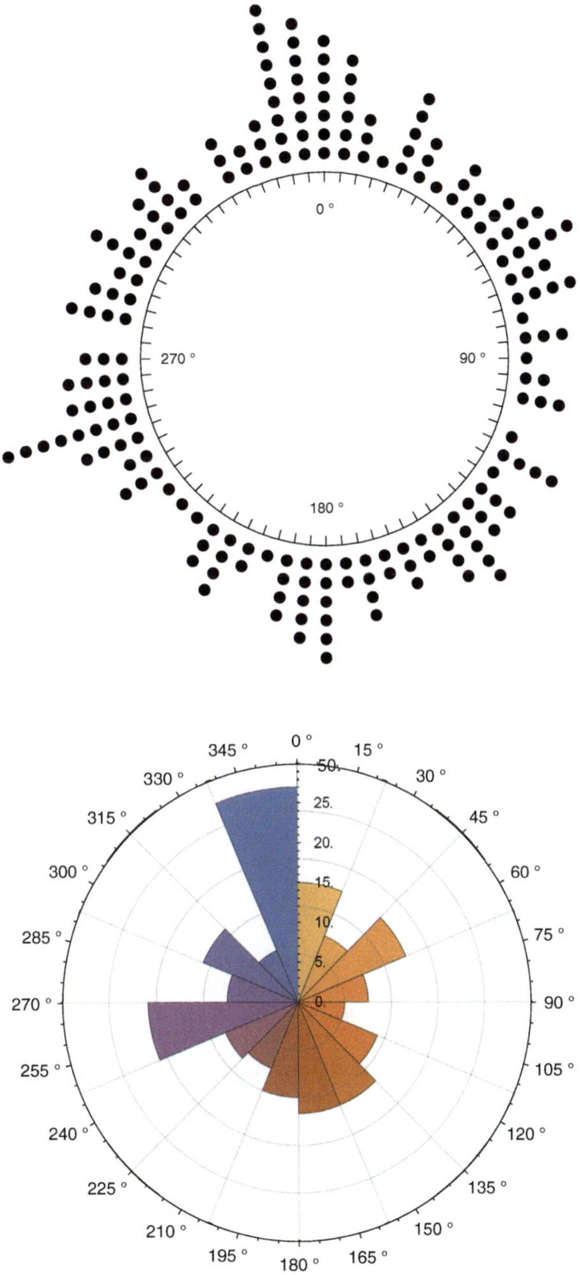

Fig. 4.8 The directions of motion obtained from the sample trajectory

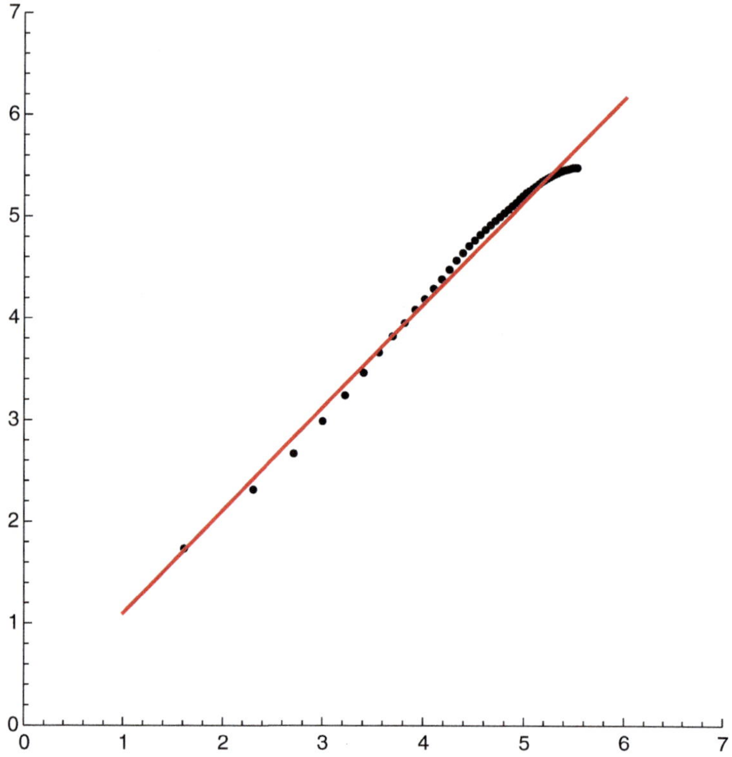

Fig. 4.9 The mean squared displacement for the sample trajectory

References

1. D. Campos, Méndez, V., Llopis, I.: Persistent random motion: uncovering cell migration dynamics. J. Theor. Biol. **267**, 526–534 (2010)
2. Dieterich, P., Klages, R., Preuss, R., Schwab, A.: Anomalous dynamics of cell migration. Natl. Acad. Sci. **105**, 459–463 (2008)
3. Fürth, R.: Die brownsche bewegung bei berücksichtigung einer persistenz der bewegungsrichtung. mit anvendungen auf die bewegung lebender infusorien [in German]. Z. Phys. **2**, 244—256 (1920)
4. Gorelik, R., Gautreau, A.: Quantitative and unbiased analysis of directional persistence in cell migration. Nat. Protocols **9**, 1931–1943 (2014)
5. The Japan meteological agency data service website: https://www.jma.go.jp/jma/menu/menureport.html. (in Japanese)
6. Masuzzo, P., Troys, M.V., Ampe, C., Martens, L.: Taking aim at moving targets in computational cell migration. Trends Cell Biol. **26**, 88–110 (2016)
7. Peruani, F., Morelli, L.G.: Self-propelled particles with fluctuating speed and direction of motion in two dimension. Phys. Rev. Lett. **99**, 010602 (2007)
8. Selmeczi, D., Li, L., Pedersen, L., Nrrelykke, S.F., Hagedorn, P.H., Mosler, S., Larsen, N.B., Cox, E.C., Flyvbjerg, H.: Cell motility as random motion: a review. Eur. Phys. J. Topics **157**, 1–15 (2008)
9. Wu, P.H., Giri, A., Wirtz, D.: Statistical analysis of cell migration in 3d using the anisotropic persistent random walk model. Nat. Protocols **10**, 517–525 (2015)

Chapter 5
Protein Structures

Hiroki Kodama and Yoichi Nakata

5.1 Introduction

Proteins mean polypeptides or complexes of polypeptides. Each polypeptide forms a single chain of 20 types of amino acids (see Figs. 5.1 and 5.2) through peptide bond polymerization and plays very important roles in all the functions of life. For example, cells are composed of proteins. Histone that wraps DNA and makes up chromatin and enzymes used in chemical reactions in living organisms such as RNA polymerase and ribosomes are also proteins. The reason why proteins have such a variety of functions is that it is possible to form various spatial structures by hydrogen bonding between places that are separated from each other on the chain.

The sequence of the 20 amino acids that make up a polypeptide chain is called the primary structure of a protein (corresponding to the genome sequence in DNA). The order of the sequence is determined by the direction of the carboxyl group from the side of the amide group that is not used for peptide binding. A partial spatial structure such as a helix or sheet structure produced by a polypeptide chain is still called a secondary structure. The structure of a polypeptide chain determined by the spatial arrangement of multiple secondary structures is called the tertiary structure, and the structure formed by a protein multimer consisting of multiple polypeptide chains with a tertiary structure is called the quaternary structure. The function of a protein is said to be directly determined by its tertiary or quaternary structure. In

H. Kodama (✉)
WPI-AIMR, Tohoku University, Miyagi, Japan

iTHEMS, RIKEN, Saitama, Japan
e-mail: kodamahiroki@gmail.com

Y. Nakata
Isotope Science Center, The University of Tokyo, Tokyo, Japan
e-mail: ynakata@ric.u-tokyo.ac.jp

T. Tokihiro (ed.), *Mathematical Modeling for Genes to Collective Cell Dynamics*, Theoretical Biology, https://doi.org/10.1007/978-981-16-7132-6_5

Name	Formula	Hydrophobicity	Name	Formula	Hydrophobicity
Alanine	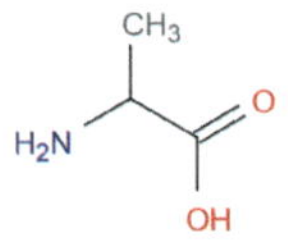	Hydrophobic	Phenylalanine	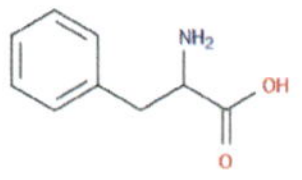	Hydrophobic
Valine	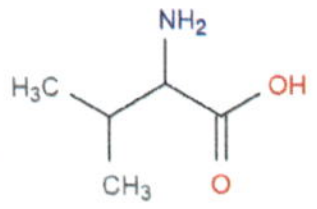	Hydrophobic	Methionine	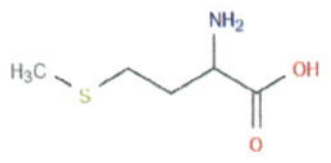	Hydrophobic
Leucine	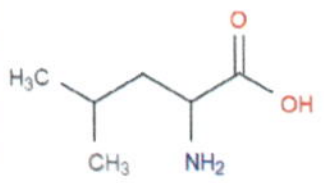	Hydrophobic	Tryptophan	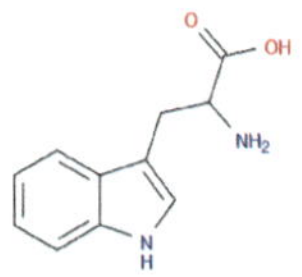	Hydrophobic
Isoleucine	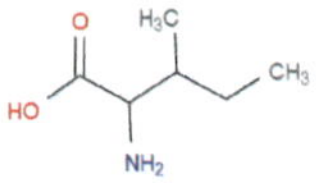	Hydrophobic	Glycine	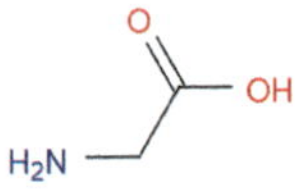	Hydrophobic
Proline	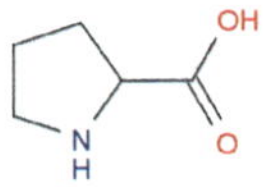	Hydropphobic	Cysteine	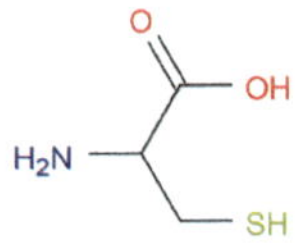	Hydrophobic

Fig. 5.1 List of the hydrophobic amino acids. The structural formulae are generated by OpenBabel [20] from data on ChemSpider [1]

this chapter, we mainly focus on the structure formed by a single polypeptide chain, as well as the secondary and tertiary structures.

It is possible to directly measure spatial structures of proteins by using X-ray, nuclear magnetic resonance, and cryo-electron microscopy, and the result of such measurements is databased as PDB [25]. CATH [4] and SCoP [26] are databases that classify the spatial structures of proteins based on PDB.

The potential of a protein is determined by the forces such as hydrogen bonding, electrostatic interactions, and van der Waals forces in the environment in which it is placed (temperature, surrounding environment, etc.). The spatial structure is considered to achieve the minimum potential. The most free energy stable structure

Name	Formula	Hydrophobicity	Name	Formula	Hydrophobicity
Lysine	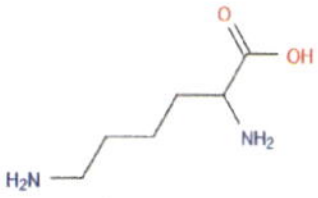	Polar	Asparagine	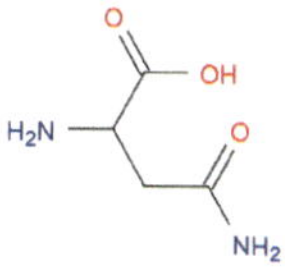	Polar
Arginine	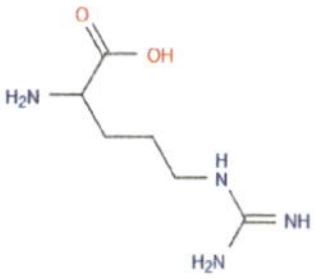	Polar	Glutamine	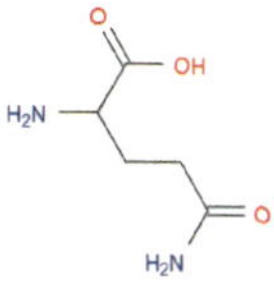	Polar
Histidine	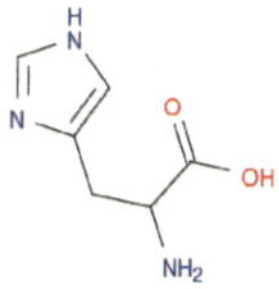	Polar	Serine	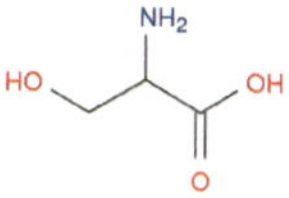	Polar
Aspartic acid	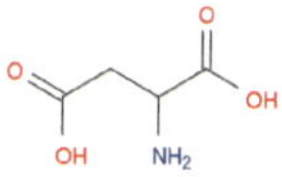	Polar	Threonine	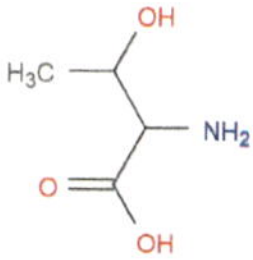	Polar
Glutamic acid	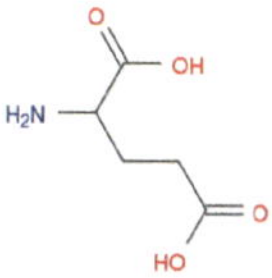	Polar	Tyrosine	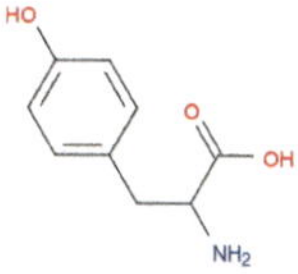	Polar

Fig. 5.2 List of the polar amino acids. The structural formulae are generated by OpenBabel [20] from data on ChemSpider [1]

that a protein takes in a cell is called a natural structure. Natural proteins have a unique natural structure.

The structure that a protein forms under the natural environment is called the natural state and that under the changed environment, for example, by changing temperature, adding acid or alkali, etc., is called the denatured state. When the causative factor is removed from the protein in the denatured state, it takes on the natural state again.

The side chains of amino acids that make up the polypeptide chain are classified into two types of affinity with water: polor (hydrophilic) and hydrophobic. Because there are many water molecules around proteins in the cell, proteins in their natural

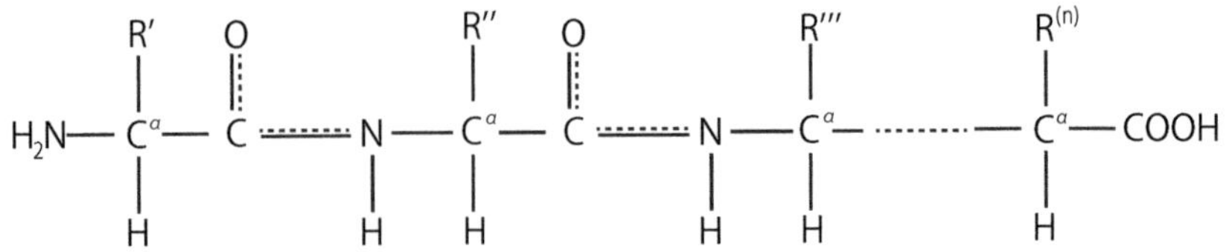

Fig. 5.3 Image of polypeptide. R, R′, . . ., $R^{(n)}$ are side chains

state are folded so that the hydrophobic side chains come to the inside and the polar side chains come to the outside.

The amino acids that make up a protein are known to take the form of the chemical structure shown in Fig. 5.3. Here, R is a radical, namely a different functional group for each amino acid, and referred to as a side chain. By calling the central carbon atom in amino acid C^α, there are four atoms bound to C^α, each of which is known to be located at the vertex of a tetrahedron centered on C^α. In addition, since the nitrogen and carbon atoms are double bonded in the peptide bond, all four atoms that are directly bonded to them exist in the same plane.

In this chapter, we will confirm that we can reproduce these phenomena in proteins using several models.

5.2 Lattice Polymer Models

5.2.1 Self-avoiding Walk

Although real polypeptide chains exist in a three-dimensional space, and there are some physical constraints such as exclusion volume effects and inter-atomic distances due to binding, there are sufficient spatial degrees of freedom for each monomeric linkage. When such a system is modeled naively, it is difficult to simulate it because the degrees of freedom are too large, and it is also very difficult to analyze the phenomena caused by the system. Therefore, we prepare a model that is easy to simulate and analyze, which is away from reality but has fewer degrees of freedom. If a toy model reflects the phenomena expected in reality, then it is natural to consider that the elements employed in constructing the model are necessary to reproduce its behavior in real systems.

As an example, let us consider the lattice polymer model called self-avoiding walk (SAW) [17], which is the simplest model to describe the structure of polymeric polymers considering the exclusion volume effect. This is the point sequence $(v_i)_{i=1}^{N}$, in which monomers are arranged on a two- or three-dimensional lattice

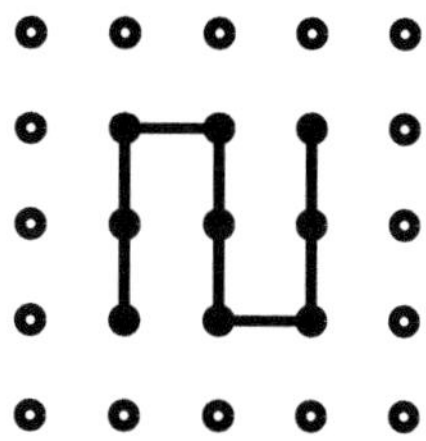

Fig. 5.4 An example of SAW

in such a way that they do not overlap (see Fig. 5.4) and mathematically satisfy the following conditions:

(i) $\mathbf{v}_i \in \mathbb{Z}^d$
(ii) $||\mathbf{v}_{i+1} - \mathbf{v}_i|| = 1$
(iii) $\mathbf{v}_i \neq \mathbf{v}_j$ for $i \neq j$

where $d \in \mathbb{Z}_{>0}$ is the dimension of the considered lattice (a planar lattice if $d = 2$ and a spatial lattice if $d = 3$), N is the length of SAW (the number of monomers that make up the molecule), and $||\mathbf{x}||$ is the Euclid norm of the vector $\mathbf{x} \in \mathbf{Z}^d$. The fact that condition (5.2.1) is a non-local condition for the monomer number i is an inherent difficulty in constructing a SAW.

Therefore, the problem of counting the number of possible SAWs under a fixed length or range of lattices is very interesting not only in polymer physics but also in mathematical physics, combinatorics [13], and algorithms [12]. A contact is defined as a pair of monomers that is not adjacent to SAW as a chain but is adjacent to it in the lattice space. That is, a contact is a set of indexes $C = \{(i, j) \mid j > i + 1, ||\mathbf{v}_i - \mathbf{v}_j|| = 1\}$.

5.2.2 *HP Model*

The HP model [6] introduces the information of polarity or hydrophobicity for each vertex (monomer) of the SAW lattice under the assumption that the other empty lattices around the polymer lattice are packed with solvent. We express simply "H" or "P" for hydrophobicity or polarity of each site, respectively. The potential of lattice protein in the HP model is calculated by -1 times the number of contacts between H sites in SAW. That is, let a path of SAW be $\mathscr{P}$, and for the contact $C = C(\mathscr{P})$ of that path, $E(C) = \sum_{(i,j)\in C} W(\sigma_i, \sigma_j)$, where $\sigma_i \in \{H, P\}$ $(i = 1, \ldots, N)$ expresses hydrophobicity or polarity of i-th site and W is $W(H, H) = -1$ and $W(H, P) = W(P, H) = W(P, P) = 0$. In other words, as $C_{\mathrm{HH}} = \{(i, j) \in C \mid \sigma_i = \sigma_j = H\} \subset C$, $E(C) = -\#C_{\mathrm{HH}}$.

As mentioned before, the structure of a protein is that minimizes the potential, and the same principle can be applied to lattice proteins. One of the main problems of the HP model is to determine the most stable structure and another is to calculate the energy expectation value.

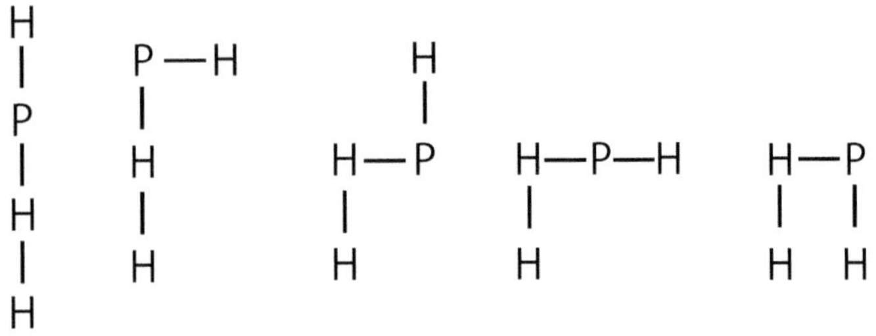

Fig. 5.5 All configuration for sequence "HHPH" under rotation and mirror symmetry in the HP model

The HP model becomes a basement of the protein folding problem from the mutations [7, 14, 15, 27, 28] to other various biological problems [16, 21, 24, 30] because it reduces the difficulty of complexity of amino acid sequences. It is also interesting for computational physics: the computation of the most stable structure of HP models in a three-dimensional lattice space is known to be NP-complete [3].

For example, in the case $d = 2$ and $N = 4$, there are four possible structures as SAWs, except for rotation and mirror images. However, by setting H or P character for each site, we can distinguish five possible structures for the HP model (see Fig. 5.5). Here, under this definition, there can be a case that two lattice proteins can be identified by inverting a sequence, e.g., "HPHH" and "HHPH" on a line. But we distinguish them as different structures because the real proteins have different terminals for each side. There are four sequences with only one natural structure: HPPH, HHHH, HHPH, and HPHH.

For a larger monomer number N, the number of possible SAWs increases explosively, so it is unrealistic to argue by enumeration. Known algorithms to compute the most stable structure for such cases include the CHCC method by Yue and Dill [31] and a method using restricted programming [18] if limited to three dimensions. If it is sufficient to obtain not the exact minimal energy, there is the approximation algorithm that can compute the structure with at least 1/4 of the minimal energy in polynomial time by Hart and Istrail [10] and 1/3 of the minimal energy by Newmann [19] for the two-dimensional model. For a three-dimensional model, a polynomial time algorithm that can compute the structure with at least 3/8 of the minimal energy by Hart and Istrail [10] is known.

5.2.3 *An Extension of HP Model: Including Coulomb Force*

Hidaka et al. are interested in how an extension of the HP model could explain the structural denaturation of proteins by pH changes from a statistical mechanistic perspective [11]. Then, they requested the HP model to be extended to suffer an energy loss when polar side chains are adjacent to each other. This is due to the fact that the polar side chains are ionized by the release of protons due to the change in pH, and the side chains have Coulomb repulsive force. That is, instead of W, we introduce a potential with $W'(H, H) = \varepsilon_{\mathrm{HH}} > 0$, $W'(P, P) = \varepsilon_{\mathrm{PP}} < 0$, $W'(H, P) = W'(P, H) = 0$ as W' and calculate the expected value of the energy $E(C) = \sum_{(i,j)\in C} W'(\sigma_i, \sigma_j) = \varepsilon_{\mathrm{HH}}\#C_{\mathrm{HH}} + \varepsilon_{\mathrm{PP}}\#C_{\mathrm{PP}}$ by counting. Note that $C_{\mathrm{PP}} := \{(i, j) \in C \mid \sigma_i = \sigma_j = P\} \subset C$.

Here, for a given length of SAW, we define t_{MAX} as the maximum contact for all considerable configurations, that is, $t_{\mathrm{MAX}} := \max_{\mathscr{P}} \#C$. For each SAW path $\mathscr{P}$, we define compactness as $\rho := \#C(\mathscr{P})/t_{\mathrm{MAX}}$, which indicates degree of spread for a path. In the case $\rho = 1$, the path is most compact, and when ρ becomes smaller, it becomes broader. Here, we note that one has $t_{\mathrm{MAX}} = N + 1 - \lceil 2\sqrt{N} \rceil$.

Assuming a Boltzmann distribution, the expectation of contact-dependent physical quantity $X = X(C)$ is denoted by $\langle X \rangle = \sum_{\mathscr{P}} X(C(\mathscr{P}))e^{-\beta E(C(\mathscr{P}))}/Z$, where $\sum_{\mathscr{P}}$ is the sum of all possible configurations for a fixed sequence, except for parallelism, rotation, and mirror symmetry, and $Z = \sum_{\mathscr{P}} e^{-\beta E(C(\mathscr{P}))}$ is the partition function.

Hidaka et al. constructed a lattice model with this potential in a two-dimensional lattice and obtained the following results.

A change of the stable state is confirmed as $\mu = -\varepsilon_{\mathrm{PP}}/\varepsilon_{\mathrm{HH}}$ grows. This means that the most stable structure is altered by pH denaturation. It was also confirmed that several new most stable structures may appear. This means that even a protein in its natural state can have multiple most stable structures after pH denaturation.

For the small length case, such a change of the stable state occurs at most once. However, it occurs twice as the length N exceeds 16. This also corresponds to multiple occurrences of pH denaturation in real proteins (see Figs. 5.6 and 5.9).

The compactness of the structure when μ is large is smaller than the structure that appears when μ is small. It is known that proteins in the denatured state take a partially folded structure (molten globule), and this phenomenon may correspond to this (Figs. 5.7, 5.8, 5.10, 5.11, and 5.12).

These results show that the planar lattice protein model is able to reproduce the properties of real proteins in spite of its simplicity.

Monte Carlo simulations are used to discuss energy expectations (it is also possible to use Monte Carlo methods to search for energy minima). In other words, the most stable structure and statistical and physical expectations are calculated by mutating the chains created by the SAWs and having them search for possible SAW configurations. However, there are several problems to execute Monte Carlo simulations. First of all, the potential of the HP model has several local minima. Then, the mutation keeps being trapped around a local minimum and cannot reach the true minimum by a naive simulation. There are two main methods to avoid this problem: one is to simulate multiple systems with different temperature parameters in parallel and sometimes to interchange the state of the systems (in fact, the temperature parameters are changed), so that multiple energy minima can be moved through the high-temperature part where the energy minima are easily overcome. This approach is called "replica exchange method" [29]. The other method is to replace the energy-based Boltzmann distribution-based displacement used in the

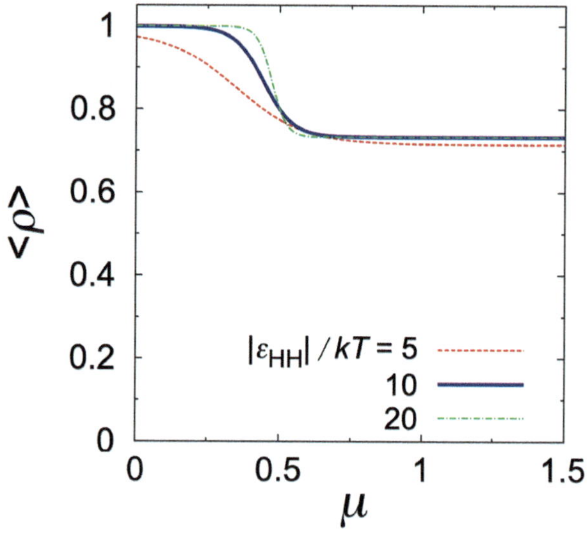

Fig. 5.6 Change of the expectation of compactness for sequence "HHPHHPPPPPHPHPPHPH." As μ grows, the stable state is changed. Here, $kT = 1/\beta$

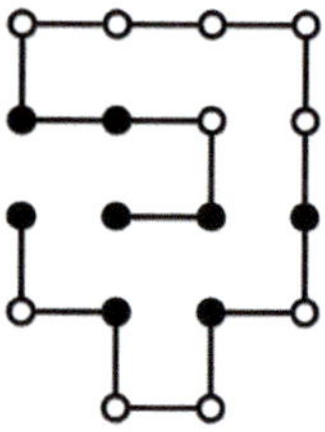

Fig. 5.7 The stable state for $\mu = 0$, which is the natural state of the lattice protein. H residues are expressed as filled circles and P as empty circles

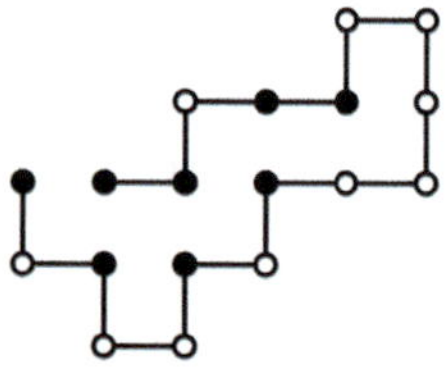

Fig. 5.8 A configuration of stable states around $\mu \sim 1$

metropolis method with another probability distribution that tends to exceed the energy minimum. When calculating the expected energy value, one resamples using Boltzmann distribution from obtained data distribution. This approach is called as "multicanonical method" [2]. For Monte Carlo simulations of general multimodality potentials, the method described above is sufficient, but for SAWs, it is very difficult

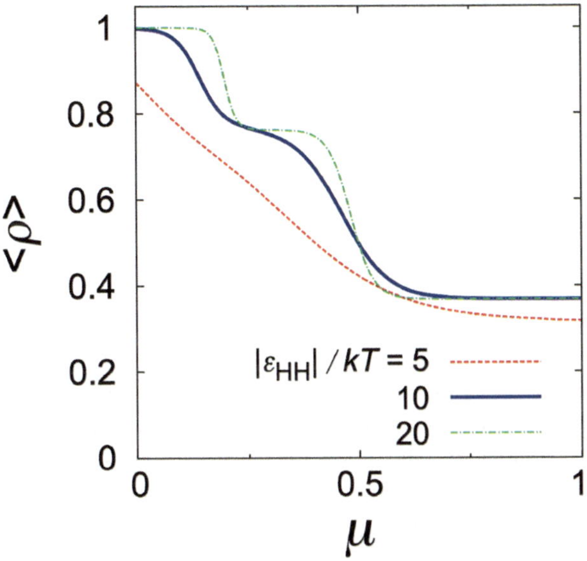

Fig. 5.9 Change of the expectation of compactness for sequence "HHPHHPPPPPHPHPPHPH"

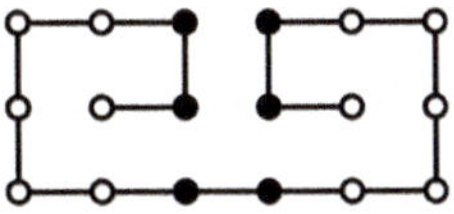

Fig. 5.10 The stable state for $\mu = 0$

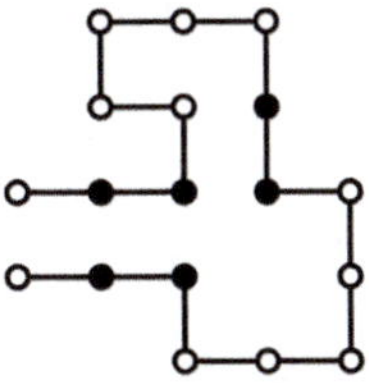

Fig. 5.11 A configuration of stable states around $\mu \sim 0.3$

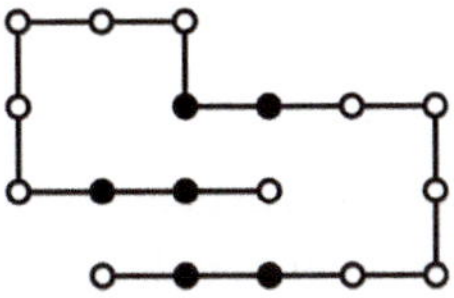

Fig. 5.12 A configuration of stable states around $\mu \sim 0.7$

to mutate preserving SAW, and it is virtually impossible to obtain a sufficient number of samples to run at least a Monte Carlo simulation. One way to avoid this is once to break the SAW rule and allow overlaps, but to select only those without overlap when resampling energy expectations [5].

5.3 Fatgraph Model

In the previous section, we discussed the structure determined by the properties of the side chains. Here, we briefly describe an abstraction model of the three-dimensional structure focusing on the hydrogen bonds formed by peptide bonds as another example of the mathematical modeling of proteins.

A peptide bond between the i-th C^{α} atom and the $i+1$-th one is called the i-th peptide unit (remember that the atoms that make up the unit are all on the same plane, see Fig. 5.13). Note that there are two types of structures that a peptide unit can take, cis and trans, and that the position of the hydrogen molecules used for hydrogen bonding is different. It should also be noted that for proline, the hydrogen molecule is on the side of the side chain, i.e., on the opposite side of the chain from the other amino acids. Unless otherwise noted, the term "C atom" refers to the C atom involved in the peptide bond in a unit.

The i-th and $i+1$-th peptide units are apparently adjacent and the hydrogen-bonded peptide units are also closely related. Note that the single bond between the i-th C^{α} atom and the C atom of the i-th peptide unit has a degree of freedom of axial

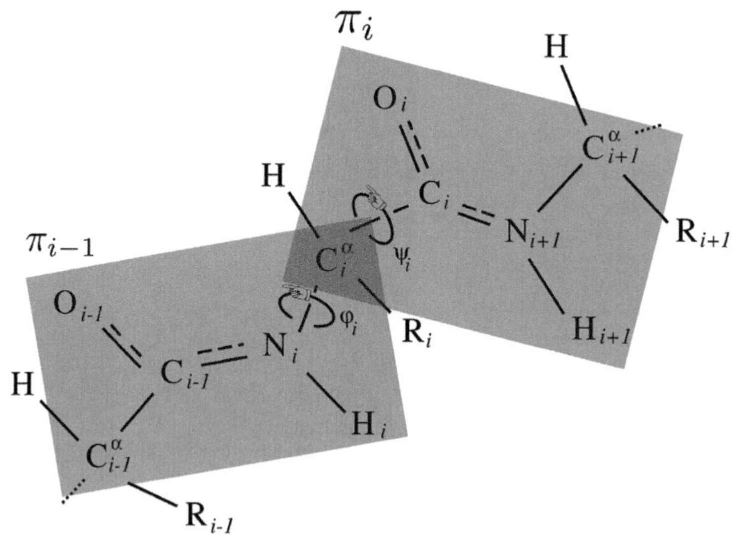

Fig. 5.13 Peptide chain and peptide unit. Arranged from [23]

rotation. The nitrogen atom and the $i+1$-th C^α atom also has a degree of freedom of axial rotation. It is immediately apparent that the rotation of the two axes creates an angle between the planes in which the two neighboring peptide units are placed. Penner et al. introduced a model that replaced the connection relationship with an extended graphical structure called FATGRAPH by a simplification that makes it binary about the angles made by this peptide unit [23]. A normal (undirected) graph is defined as an edge $E \subset \{\{i, j\} \mid i, j \in V\}$ connecting a vertex set V to a vertex, whereas a fatgraph is given two attributes, twisted or untwisted, for each edge $e \in E$. Here, we note that one can introduce a cyclic order for the edges starting from a vertex. A fatgraph is a displacement retract of a surface that uniquely defines a surface with a fatgraph as its boundary. We denote π_i as the plane on which i-th peptide unit is on. Now, let us construct an orthonormal system (u_i, v_i, w_i) on $\mathbb{R}^3$ from i-th peptide unit by following rules:

- u_i is a normalized vector from the C atom to the N atom on π_i.
- v_i is a unit vector on π_i which is orthogonal to u_i and close to the O atom in view of the C^α atom.
- $w_i = u_i \times v_i$.

By regarding orthonormal system $F_i = (u_i, v_i, w_i)$ as an element of special orthogonal group SO(3), $F_i \in$ SO(3) is called a frame. Next, let us consider the connection relation between hydrogen bonding units. Given a fixed coordinate system, one can define the transformation matrix $\hat{R}_{ij}$ from F_i to F_j by $\hat{R}_{ij} F_i = F_j$, thus $\hat{R}_{ij} = F_j F_i^{-1}$. Then, the same transformation considering on the basis F_i is expressed as $R_{ij} := F_i^{-1} \hat{R}_{ij} F_i = F_i^{-1} F_j$. This is the transformation matrix from the frame of i-th peptide unit to that of j-th one when viewing from the coordinate system of the i-th peptide unit, which represents the torsion between the two peptide units. Here, we introduce two matrices $I = \begin{bmatrix} 1 & 0 & 0 \\ 0 & 1 & 0 \\ 0 & 0 & 1 \end{bmatrix}$ and $J = \begin{bmatrix} 1 & 0 & 0 \\ 0 & -1 & 0 \\ 0 & 0 & -1 \end{bmatrix}$, which are elements of $SO(3)$ and expressing two frames are untwisted or twisted, respectively (however, if the unit is trans type or $i+1$-th amino acid is proline, the relation is inverted, and if both are satisfied, the relation becomes normal). We binarize the peptide unit connections according to whether the transformation matrix R_{ij} is close to I or J. SO(3) defines $d(A, B) = \arccos(\frac{\mathrm{tr}(AB^{-1})-1}{2})$ as the natural distance. Using this, let $d(R_{ij}, I) < d(R_{ij}, J)$, then the conjunctive relation between i and j is untwisted, and let $d(R_{ij}, I) > d(R_{ij}, J)$, then the conjunctive relation between i and j is twisted. Note that if $R = (r_{k,l})_{1 \le k,l \le 3}$, then $d(R, I) < d(R, J) \Leftrightarrow r_{2,2} + r_{3,3} > 0$.

It is known that the fatgraphs of polypeptides by this definition from the two polypeptides are consistent, and then the spatial structure of the original protein is also roughly the same [23].

Therefore, despite the extreme simplicity of the polypeptide connection, fatgraph reproduces the structure of the polypeptide to some extent, and in this sense, it is a good model. For example, the famous α-Helix and β-sheet structures, which are

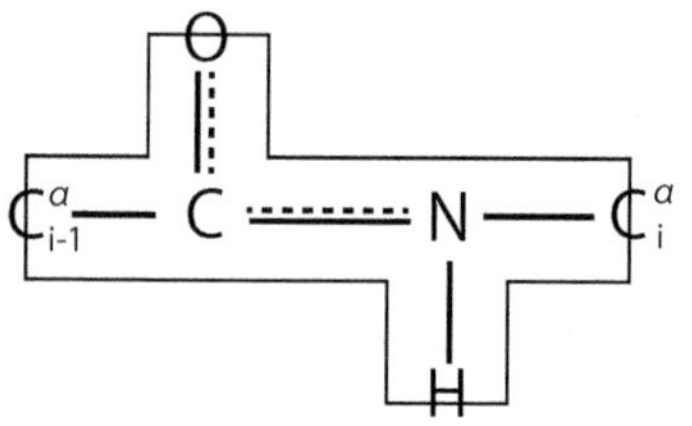

Fig. 5.14 Simple representation of peptide unit to fatgraph model. Each unit is expressed as a ribbon with branches. Ribbon itself has a symmetry under 180° rotation. However, by considering the structure of amino acids and peptide bonds, the structure is determined uniquely under setting initial configuration for a site

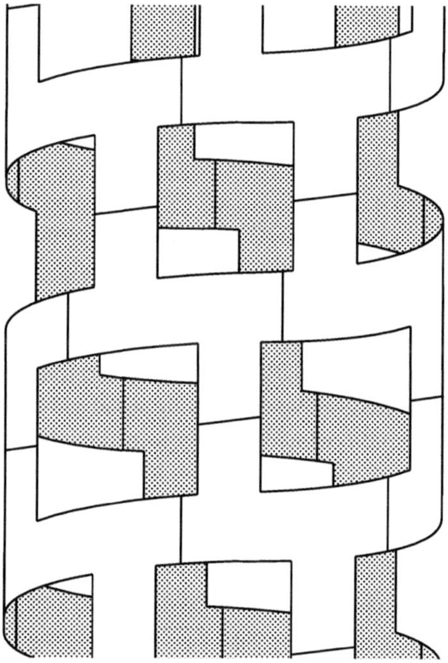

Fig. 5.15 α-Helix structure with ribbons. Reverse side of ribbons is expressed as dark

the secondary structures of proteins, are realized in the fatgraph. By expressing a peptide unit as a ribbon, these second structures are shown in Figs. 5.15 and 5.16.

Fatgraph represents the broad three-dimensional structure of the protein well but loses too much information to recapture local properties such as how strongly it binds to the ligand. We defined a transformation matrix R_{ij} belonging to SO(3), but we dropped the information by binarizing it to drop it into a fatgraph. Therefore, one approach is to adopt a model that considers the transformation matrix as it is in SO(3) [22]. Penner, Andersen et al. computed the distribution of the transformation

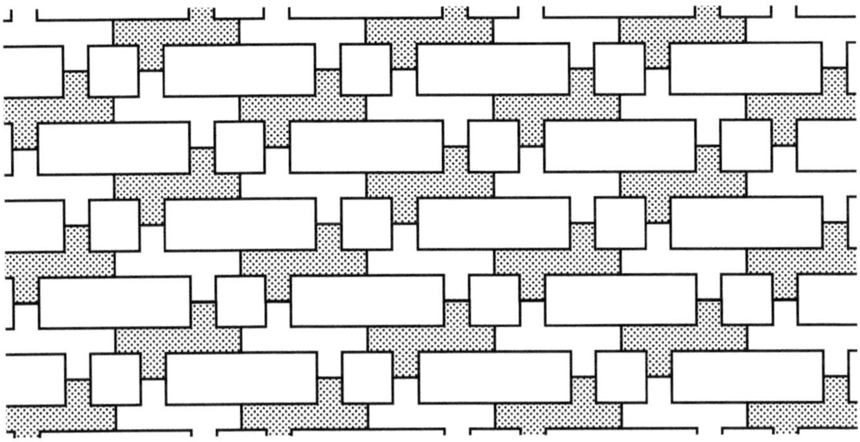

Fig. 5.16 β-Sheet structure with ribbons

matrix in SO(3) for each protein and predicted the structure of the protein based on the distribution and got good results.

We can approximate the spatial structure of a protein by the union of balls K whose centers are on the three-dimensional positions of atoms, respectively, and discuss its structure as a subset in the three-dimensional spaces, for example, the number of holes or that of cavity, which is an element of the homology basis. Furthermore, by increasing the radii of balls as time evolves, if each radius is sufficiently small, K is just a disjoint union of balls (the number of connected components is equal to that of atoms). As the radii become larger, some of the balls overlap (the number of connected components becomes smaller), and holes (1-cycle) and cavities (2-cycle) appear. When the radii become larger, some of such cycles disappear, and the set K finally becomes that which homeomorphic to one ball. The persistent homology is suggested to express the survive time of cycles in this growth [9]. An application of the persistent homology to discussion for protein structures is presented in [8]. The homology itself expresses the character of a set without quantitative discussion. However, via the time growth of radii, the persistent homology can discuss the size of cycles quantitatively in a way.

Note that none of the models directly addresses protein changes. An important question is how a protein, which is linear as an initial nutrient, can obtain a stable structure as described above when the protein is synthesized from amino acids.

References

1. http://chemspider.com/
2. Berg, B., Neuhaus, T.: Multicanonical ensemble: a new approach to simulate first-order phase transitions. Phys. Rev. Lett. **68**(1), 9–12 (1992)

3. Berger, B., Leighton, T.: Protein folding in the hydrophobic-hydrophilic (HP) model is NP-complete. J. Comput. Biol. **5**(1), 27–40 (1998)
4. Cath protein structure classification database. http://www.cathdb.info
5. Chikenji, G., Kikuchi, M., Iba, Y.: Multi-self-overlap ensemble for protein folding: ground state search and thermodynamics. Phys. Rev. Lett. **83**(9), 1886–1889 (1999)
6. Dill, K.A.: Theory for the folding and stability of globular proteins. Biochemistry **24**(6), 1501–1509 (1985)
7. Dill, K.A., Bromberg, S., Yue, K., Chan, H.S., Fiebig, K.M., Yee, D.P., Thomas, P.D.: Principles of protein folding–a perspective from simple exact models. Protein Sci. **4**, 561–602 (1995)
8. Gameiro, M., Hiraoka, Y., Izumi, S., Kramar, M., Mischaikow, K., Nanda, V.: Topological measurement of protein compressibility via persistence diagrams. Jpn. J. Ind. Appl. Math. **32**(1), 1–17 (2014)
9. Gunna, C.: Topology and data. AMS Bull. **2**, 255–308 (46)
10. Hart, W., Istrail, S.: Fast protein folding in the hydrophobic-hydrophillic model within three-eights of optimal. J. Comput. Biol. **3**(1), 53–96 (1996)
11. Hidaka, T., Shimada, A., Nakata, Y., Kodama, H., Kurihara, H., Tokihiro, T., Ihara, S.: Simple model of pH-induced protein denaturation. Phys. Rev. E **92**, 012709 (2015)
12. Iwashita, H., Kawahara, J., Minato, S.: ZDD-based computation of the number of paths in a graph. Tech. rep., Hokkaido University, Division of Computer Science, TCS Technical Reports (2012)
13. Jensen, I.: Enumeration of compact self-avoiding walks. Comput. Phys. Commun **142**, 109–113 (2001)
14. Lau, K.F., Dill, K.A.: A lattice statistical mechanics model of the conformational and sequence spaces of proteins. Macromolecules **22**, 3986–3997 (1989)
15. Li, H., Helling, R., Tang, C., Wingreen, N.: Emergence of preferred structures in a simple model of protein folding. Science **273**, 666–669 (1996)
16. Li, Y.W., Wüst, T., Landau, D.P.: Generic folding and transition hierarchies for surface adsorption of hydrophobic-polar lattice model proteins. Phys. Rev. E **87**, 012706 (2013)
17. Madras, N., Slade, G.: The Self-Avoiding Walk. Birkhäuser, Boston (1993)
18. Mann, M., Will, S., Backofen, R.: CPSP-tools—exact and complete algorithms for high-throughput 3D lattice protein studies. BMC Bioinformatics **9**, 230 (2008)
19. Newmann, A.: A new algorithm for protein folding in the hp model. In: Proceedings of the Thirteenth Annual ACM-SIAM Symposium on Discrete Algorithms, pp. 876–884 (2002)
20. Openbabel. http://openbabel.org/
21. Pattanasiri, B., Li, Y.W., Landau, D.P., Wüst, T., Triampo, W.: Conformational transitions of a confined lattice protein: a Wang-Landau study. J. Phys. Conf. Ser. **402**, 012048 (2012)
22. Penner, R.C., Andersen, E.S., Jensen, J.L., Kantcheva, A.K., Bublitz, M., Nissen, P., Rasmussen, A.M.H., Svane, K.L., Hammer, B., Rezazadegan, R., Nielsen, N.C., Nielsen, J.T., Andersen, J.E.: Hydrogen bond rotations as a uniform structural tool for analyzing protein architecture. Nat. Commun. **5**, 5803 (2014)
23. Penner, R.C., Knudsen, M., Wiuf, C., Andersen, J.E.: Fatgraph models of proteins. Commun. Pur. Appl. Math. **63**, 1249–1297 (2010)
24. Ping, G., Yuan, J.M., Vallieres, M., Dong, H., Sun, Z., Wei, Y., Li, F.Y., Lin, S.H.: Effects of confinement on protein folding and protein stability. J. Chem. Phys. **118**, 8042 (2003)
25. Protein database. https://www.rcsb.org
26. Scop: Structural classification of proteins. http://scop.mrc-lmb.cam.ac.uk
27. Shi, G., Vogel, T., Wüst, T., Li, Y.W., Landau, D.P.: Effect of single-site mutations on hydrophobic-polar lattice proteins. Phys. Rev. E **90**, 033307 (2014)
28. Shotle, D., Chan, H.S., Dill, K.A.: Modeling the effects of mutations on the denatured states of proteins. Protein Sci. **1**, 201–215 (1992)
29. Swendsen, R.H., Wang, J.S.: Replica Monte Carlo simulation of spin glasses. Phys. Rev. Lett. **57**, 2607–2609 (1986)

30. Swetnam, A., Allen, M.P.: Selective adsorption of lattice peptides on patterned surfaces. Phys. Rev. E **85**, 062901 (2012)
31. Yue, K., Fiebig, K.M., Thomas, P.D., Chan, H.S., Shakhnovich, E.I., Dill, K.A.: A test of lattice protein folding algorithms. Proc. Natl. Acad. Sci. **92**(1), 325–329 (1995)

GPSR Compliance

The European Union's (EU) General Product Safety Regulation (GPSR) is a set of rules that requires consumer products to be safe and our obligations to ensure this.

If you have any concerns about our products, you can contact us on ProductSafety@springernature.com

In case Publisher is established outside the EU, the EU authorized representative is:

Springer Nature Customer Service Center GmbH
Europaplatz 3
69115 Heidelberg, Germany

Batch number: 10726060

Printed by Printforce, the Netherlands